Geological and Geophysical Evidence for Deep Subduction of Continental Crust Beneath the Pamir

V. S. Burtman
Geological Institute
Russian Academy of Sciences
109017 Moscow, Russia

and

Peter Molnar
Department of Earth, Atmospheric, and Planetary Sciences
Massachusetts Institute of Technology
Cambridge, Massachusetts 02139

THE GEOLOGICAL SOCIETY OF AMERICA 1888

SPECIAL PAPER

281

1993

Published by The Geological Society of America, Inc.
3300 Penrose Place, P.O. Box 9140, Boulder, Colorado 80301

Printed in U.S.A.

GSA Books Science Editor Richard A. Hoppin

Library of Congress Cataloging-in-Publication Data

Burtman, V. S. (Valentin Semenovich)
Geological and geophysical evidence for deep subduction of continental crust beneath the Pamir / V.S. Burtman and Peter Molnar.
p. cm.—(Special paper ; 281)
Includes bibliographical references.
ISBN 0-8137-2281-0
1. Geology, Structural—Pamir Region. 2. Subduction zones—Pamir Region. I. Molnar, Peter Hale. II. Title. III. Series: Special papers (Geological Society of America) ; 281.
QE634.P36B87 1993
551.8'095496—dc20 93-17404
CIP

Cover photo: View northwest of folded Jurassic and Cretaceous rock of the Southeastern Pamir, from China looking toward Tadjikistan. The dark red rocks in the axis of the syncline are Cretaceous-Paleogene redbeds. They overlie the well-bedded Jurassic limestone of the Aghil Formation that dips to the left in the lower right of the photo. Triassic shale and limestone, also of the Aghil Formation, form the mountains in the background. Photograph by P. Molnar (June 1992).

10 9 8 7 6 5 4 3 2 1

Contents

Acknowledgments

We thank M. W. Hamburger for a very thorough review of the manuscript and for providing us with original copies of Figures 17 and 23 and a preprint of his work in advance of publication. J. Faulds also offered much valuable constructive criticism of this manuscript. M. R. Nelson digitized the seismograms for the earthquakes in August 1974 and carried out some preliminary analysis of them. W. Leith, V. G. Trifonov, and R. L. Wesson provided us with originals of Figures 22, 28, and 26, respectively. We benefited from discussions with Ya. A. Bekker, V. I. Dronov, T. G. Guseva, M. W. Hamburger, V. I. Khalturin, V. K. Kuchai, E. Ya. Leven, A. A. Lukk, I. L. Nersesov, I. I. Pospelov, T. G. Rautian, S. V. Ruzhentsev, S. F. Skobelev, V. G. Trifonov, and R. L. Wesson. We are particularly grateful to Zheng Hu and P. Tapponnier for inviting us to attend and supporting our attendance at the International Symposium on the Kunlun and Karakorum Mountains in June 1992 in Kashgar, China; to N. Arnaud, M. Brunel, Chen Yang, J.-P. Cogné, Liu Qing, Pan Yusheng, Sun Dongli, P. Tapponnier, and Xu Ronghua for discussing their results in advance of publication; and to P. Choukroune, A. Khan, M. Mattauer, J. Mercier, M. P. Searle, E. Sobel, A. Steck, and R.A.K. Takirkheli for their insights during the postconference field trip. This work was begun with support from the National Science Foundation under grant EAR-890476, but completion of it required support from the National Aeronautical and Space Administration under grant NAG5-795 and the Academy of Sciences of the U.S.S.R. (now Russia) as well as the patience and indirect support of others. The Centre National de la Recherche Scientifique of France and Academia Sinica of China financed our participation in the International Symposium on the Kunlun and Karakorum Mountains.

Geological Society of America
Special Paper 281
1993

Geological and Geophysical Evidence for Deep Subduction of Continental Crust Beneath the Pamir

ABSTRACT

Geological and geophysical observations imply Cenozoic subduction of intact Eurasian continental lithosphere, approximately 300 km in downdip length and including relatively thin (20–25 km) continental crust, beneath the Pamir. An inclined seismic zone dips at about 45° south-southeastward to a depth of 150 to 200 km beneath the Pamir and projects to the surface near the northern margin of the Pamir. The downdip length of the seismic zone of about 300 km implies a comparable amount of subduction of lithosphere in late Cenozoic time. The seismicity and tectonically most active part of the Pamir and its surroundings follows the northern margin of the Pamir. Quaternary offsets on faults and repeated geodetic observations suggest that roughly half of India's present 44-mm/a convergence with Eurasia is absorbed by localized crustal shortening and underthrusting at this zone.

Three kinds of geologic observations suggest that the outer (northern) margin of the Pamir has been thrust at least 300 km over southern Eurasia. A Paleozoic suture between Eurasia and continental fragments originally from Gondwanaland crosses northern Afghanistan and follows the Kunlun in northern Tibet. Between Afghanistan and Tibet it is deflected roughly 300 km northward around the Northern Pamir. Large systematic anomalies in paleomagnetic declinations measured in Cretaceous and Paleogene sedimentary rock on the northern and western margins of the Pamir can be explained by bending of this margin and by 300 to perhaps 700 km of northward displacement of the Pamir with respect to the region farther west and north. East-west–trending facies boundaries in Cretaceous and Paleogene sedimentary rock in the Tadjik Depression are abruptly truncated at the western edge of the Pamir. Their eastward continuations, which follow the northern margin of the Pamir, lie a minimum of 200 km farther north. In addition, crustal shortening within the outer zone of the Pamir appears to exceed 100 km. Together this displacement of 200 km and the internal shortening of about 100 km also imply that the Northern Pamir has been displaced northward 300 km or more with respect to the rest of Eurasia farther north.

If continental crust has been subducted into the asthenosphere beneath the Northern Pamir, some other process must account for the present thick crust beneath the Pamir. Geologic mapping of major thrust and strike-slip faults within the Pamir indicates that crustal shortening, largely by thrust faulting, has absorbed more than 300 km of north-south convergence. Hence the southern part of the Pamir has been displaced northward this additional amount with respect to Eurasia. Crustal thickening within the Pamir, from normal thickness to roughly 70 km, can account for about 300 ± 100 km of convergence and therefore can account for the Cenozoic shortening within the Pamir. In contrast, the crust presently beneath the Tadjik Depression, west of the Pamir, includes a relatively thick sequence of Mesozoic and Cenozoic deposits (10–15 km) overlying a relatively thin crystalline basement (20–25 km). The sedimentary cover the Tadjik Depression has undergone folding and faulting in the Cenozoic era; the basement of most of the depression appears to have remained intact.

Burtman, V. S., and Molnar, P., 1993, Geological and Geophysical Evidence for Deep Subduction of Continental Crust Beneath the Pamir: Boulder, Colorado, Geological Society of America Special Paper 281.

The geography and estimates of amounts of convergence that can be accounted for in different ways imply that the intermediate-depth earthquakes beneath the Pamir occur within the lithosphere that lay east of the Tadjik Depression before it was overridden by the Pamir. We assume that the thin crystalline basement that presently underlies the Tadjik Depression characterizes that subducted lithosphere. Its sedimentary cover was detached, and some of it now crops out along the northern margin of the Pamir, but the continental lithosphere with its 20 to 25 km of crystalline crustal basement has been subducted intact to a depth of 200 km or more. The present rate of convergence across the Pamir suggests that this process continues unabated.

INTRODUCTION

Subduction of lithosphere along narrow thrust zones and to great depths in the asthenosphere under continental regions appears to be rare. It it the most poorly represented among those processes that combine to make plate tectonics the simple and elegant description of the tectonics of oceanic regions. Continental rift systems are an obvious analogue for oceanic spreading centers. Most major strike-slip faults can be viewed, at least qualitatively, as transform faults. Convergence between continental regions, however, is rarely, if ever, as simple and localized as underthrusting of oceanic lithosphere at subduction zones. Instead, deformation is commonly spread over a broad region. and the basic rules of plate tectonics do not seem to provide much of a guide toward understanding that deformation. Where studied, as in central and eastern Asia the diffuse deformation between continents seems to absorb intracontinental convergence without subduction of continental crust either to great depth or as a steady process (e.g., Molnar and Tapponnier, 1975; Molnar and others, 1987; Tapponnier and others, 1986).

If, nevertheless, continental crust could be subducted into the asthenosphere, our understanding of several processes at the forefront of the earth sciences would be affected. The budget of crustal material involved in continental collision would obviously be incomplete. Estimates of shortening across orogenic belts might vary markedly along strike, between regions where peninsulas first collide and those where embayments meet later. The exposed geologic history of mountain belts not only would provide inaccurate qualitative descriptions of the processes that occurred but also might yield deceptive images of the processes at depth and particularly those dynamic processes that govern the overall rates and extent of deformation. In addition, subduction of continental crust would contaminate the chemistry of the mantle with intense doses of radiogenic nucleides, incompatible elements, minerals that melt a low temperatures, and possibly water. Some continental crust might melt locally and then simply underplate the overlying crust. Other subducted crust might be swept into the convective circulation in the mantle and contribute to its lateral heterogeneity. A small amount of continental crust might account for what is sometimes attributed to detached and remelted subcontinental lithosphere (e.g., McKenzie and O'Nions, 1983) or to primitive mantle. Thus, quantifying the amount of continental crust that is, or can be, subducted is a worthy endeavor, even if the amount is small.

Simple calculations of the temperature structure, and hence of the density structure, of lithosphere show that a full complement of normal continental lithosphere, with 35 km of crust, should be too buoyant to undergo steady state subduction (McKenzie, 1969; Molnar and Gray, 1979). Continental crust as thin as 10 km, however, should not be too buoyant to prevent such subduction. Given the uncertainties in all of the relevant parameters, plus the possibility of phase changes to denser minerals in the crust (Richardson and England, 1979), we cannot eliminate the possibility of deep subduction of continental crust 20 km thick (Molnar and Gray, 1979). The latter possibility is especially plausible if such crust is attached to old oceanic lithosphere, such as peninsulas or microcontinents on the trailing edge of a downgoing slab of oceanic lithosphere. Moreover, if upper crust were easily detached from lower crust along gently dipping, mid-crustal shear zones, subduction of the lower crust and the underlying continental mantle lithosphere could be self-sustaining (Fleitout and Froidevaux, 1982; Molnar and Gray, 1979). Thus, we have no grounds for presuming that continental crust never is subducted.

Subduction of continental crust seem to have occurred locally. Roecker's (1982) observation of low seismic wave velocities surrounding the intermediate-depth seismic zone beneath the Hindu Kush is puzzling, unless continental crust was subducted. In contrast, high velocities commonly characterize inclined seismic zones associated with subduction of oceanic lithosphere. Earthquakes and high velocities presumably occur within the downgoing slabs because the slab is cold and strong enough to sustain the stresses needed to cause earthquakes (e.g., Isacks and others, 1968). Low velocities in such settings are most easily explained by subduction of low-velocity crust. Roecker and others (1987) observed a similar low-velocity zone well below the Moho under Taiwan, suggesting subduction of Asian continental crust there. More convincing evidence for subduction of continental crust is provided by Chopin's (1987) discovery of gneisses, from the western Alps, containing minerals subjected to pressured of 3 GPa (depths of about 100 km). These studies thus demonstrate that continental crust is too buoyant to be subducted

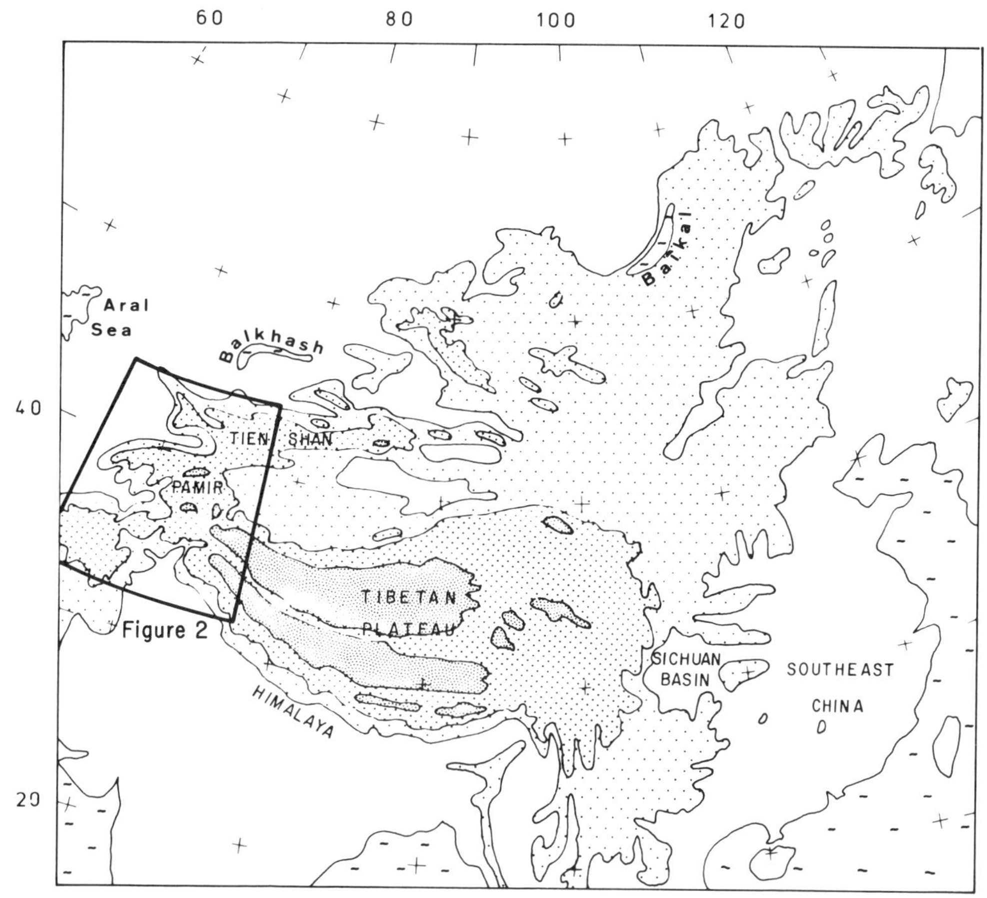

Figure 1. Regional map of Asia showing topography and the area discussed here (Figs. 2 and 24). Shaded areas, in decreasing order of darkness, show areas 5,000 m and higher, 3,000 m up to 5,000 m, and 1,000 m up to 3,000 m. White areas are lower than 1,000 m. Dashed regions show large bodies of water (oceans and large lakes).

into the asthenosphere, but they do not require it to be a major process.

One might hope that accurate geologic mapping across orogenic belts could constrain the amount of continental lithosphere subducted into the asthenosphere, but published estimates are controversial. For instance, several Alpine geologists have concluded that a few hundred kilometers of continental crust of normal thickness have been subducted beneath the Alps (Butler, 1986; Butler and others, 1986; Grandjacquet and LePichon, 1987; Laubscher, 1988; LePichon and others, 1988). Other researchers, recognizing that the crust of Eurasia's ancient southern margin thinned when that margin formed, have deduced that such subduction is not demanded by the structure and geologic history of rock exposed in the Alps (Helwig, 19796; Ménard and others, 1991; Platt, 1986). These studies leave open the possibility that continental crust has been subducted beneath the Alps, but none requires such a process.

Our purpose here is to constrain the rate and amount of crustal subduction beneath the Pamir, the high terrain that is the northwestern topographic prolongation of the Tibetan Plateau (Figs. 1 and 2). Even though the geologic history of the Pamir is not known in nearly the same detail as that of the Alps, the Pamir offers several features that are absent in the Alps but that are much more useful for addressing crustal subduction than detailed surface geology. We do not present a detailed description of the geologic history of the Pamir Instead this paper is written in modules, with each summarizing evidence that provides a quantitative bound on some aspect of the budget of crustal subduction in the Pamir.

GEOLOGICAL CONSTRAINTS ON THE MAGNITUDE OF CONVERGENCE BETWEEN THE PAMIR AND THE REST OF EURASIA

Geological evidence of three types shows that the outer (northern) arc of the Pamir has been thrust hundreds of kilometers northward with respect to the South Tien Shan and the rest of Eurasia. First, ancient orogenic belts crossing Afghanistan and Tibet appear to be deflected northward 300 km in the

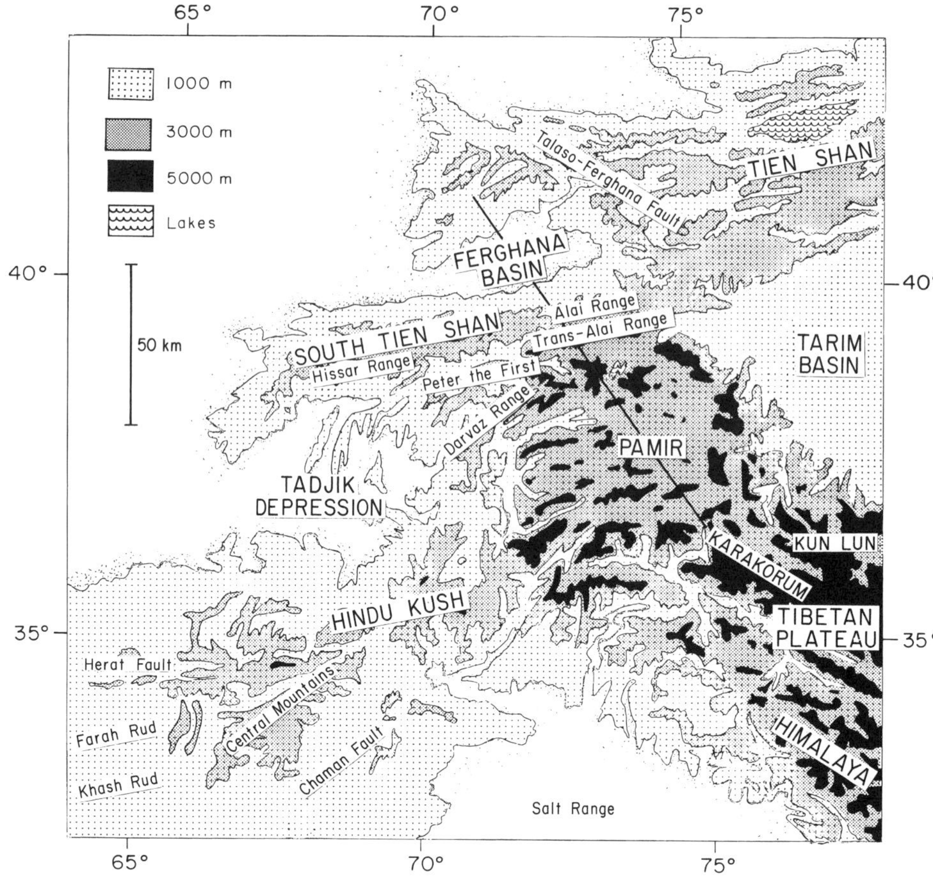

Figure 2. Simplified contour map of the Pamir and surrounding regions showing areas of interest to this study and principal topographic features. Dark black line shows the axis of the cross section of seismicity in Figure 35.

Pamir. Second, paleomagnetic declinations from Cretaceous and Paleogene sedimentary rock show large rotations that suggest substantial Cenozoic bending of the outer arc of the Pamir and between 300 and 700 km of northward displacement of it with respect to the South Tien Shan. Third, truncations and displacements of facies belts in Cretaceous and Paleogene sedimentary rock from the Tadjik Depression and Pamir-Alai Range imply a *minimum* of 200 km of northward displacement of the sedimentary rock of the outer arc of the Pamir with respect to their continuations farther west. Intense shortening of more than 100 km of this sedimentary rock implies a total of more than 300 km of convergence. In addition, some 350 km or more of shortening within the Pamir appears to have occurred in Cenozoic time.

Deflections of Paleozoic and Mesozoic sutures in the Pamir

Many of the grand thinkers of Asian geology, such as Argand (1924), Hayden (1915), Nalivkin (1915, 1926), Suess (1904, pp. 439–447, 459–460), and Wadia (1975, pp. 401–402), recognized that southern Asia developed with successive deformation of east-west–trending belts that pass through Afganistan, wrap around the Pamir, and cross Tibet (Figs. 3 and 4). Burtman and others (1963), Peive and others (1964), and Zakharov (1964) showed that the marked deflections of these belts around the Pamir require at least 250 to 300 km of northward displacement of the Pamir relative to the eastward continuations of the belts and therefore some 250 to 300 km of right-lateral strike-slip displacement, or shear, along the eastern margin of the Pamir. Others have extended these inferences to include comparable amounts of left-lateral displacement along the western margin of the Pamir (e.g., Legler and Przhiyalgovskaya, 1979, Tapponnier and others, 1981; Zakharov, 1964).

Deformation in the main tectonic belts of the Pamir occurred in late Paleozoic ("Hercynian" or "Variscan"), late Mesozoic ("Cimmerian"), and Cenozoic ("Alpine") time. With the recognition that small continental fragments could be carried long distances by oceanic lithosphere, these belts are now associated with successive collisions of fragments of crust with southern Eurasia and with the closing of three roughly east-west–trending ocean basins (Fig. 3) (e.g., Bassoulet and others,

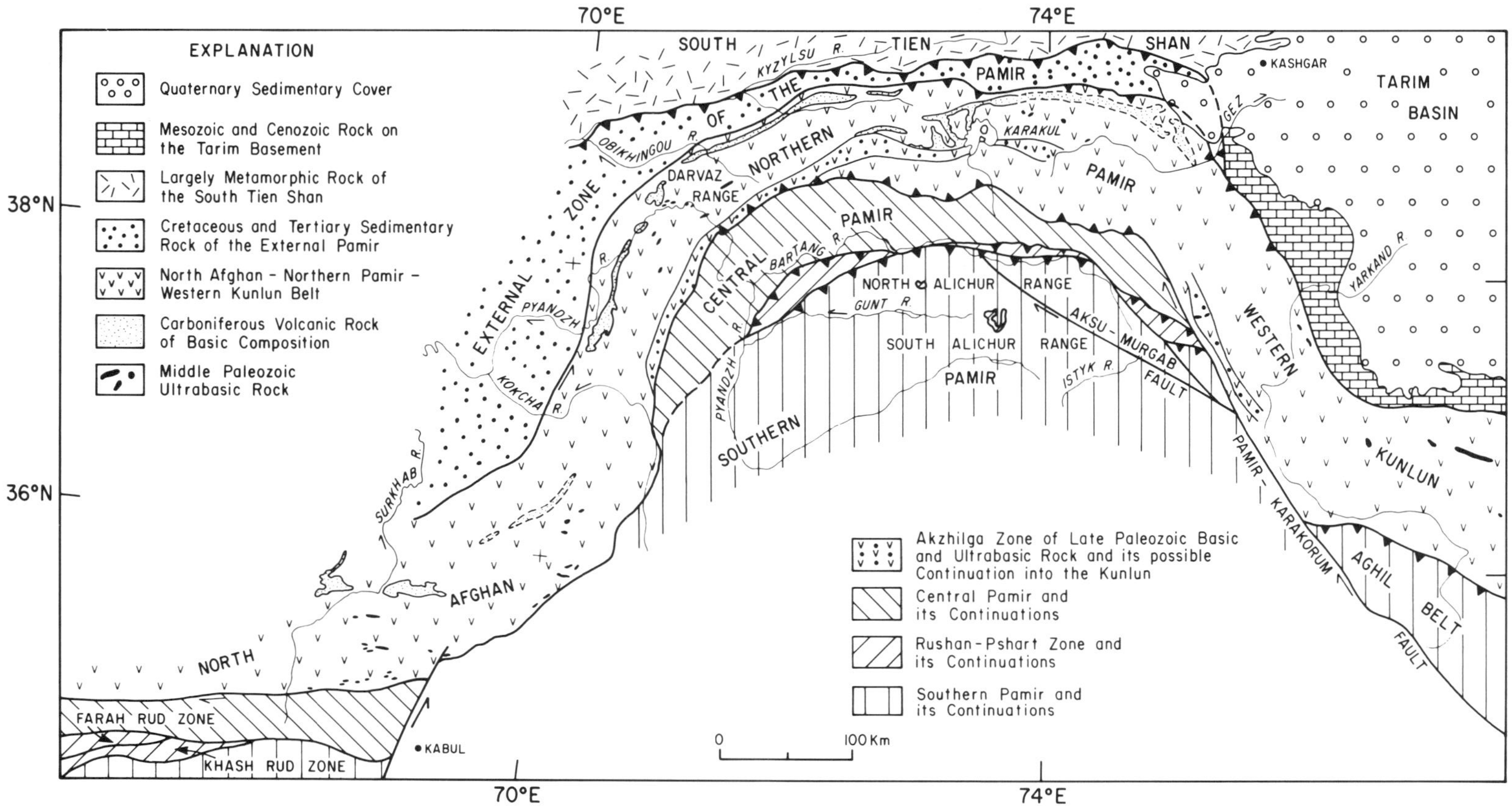

Figure 3. Map of simplified geology showing major tectonic units in the Pamir region, including the Paleozoic and Mesozoic sutures and remnants of oceanic crust. This has been adapted from the Geological map of Afghanistan (1977), the Geological map of the Tadjik SSR and adjoining territories (1984), the Geological map of the Xinjiang-Uyghur Autonomous Region, China (1985), and the Geological map of the Tibetan Plateau (1980) and from data of Budanov and Pashkov (1988), Kazmin and Faradzhev (1961), and Sinitsyn (1957). Notice that a late Paleozoic belt that includes middle to late Paleozoic volcanic and ultramafic rock wraps around the Northern Pamir, from northern Afghanistan and into the western Kunlun. A second possible suture zone is marked by the Rushan-Pshart zone within the interior of the Pamir and its western equivalents, the Farah Rud and Khash Rud zones. The eastern equivalent lies east of the area shown on this map. Small klippen of ophiolitic rock from the Rushan-Pshart zone lie within the north and south Alichur ranges. Heavy lines with teeth mark major thrust faults, with teeth pointing in down-dip directions. Heavy lines without teeth indicate active, or recently active, strike-slip faults.

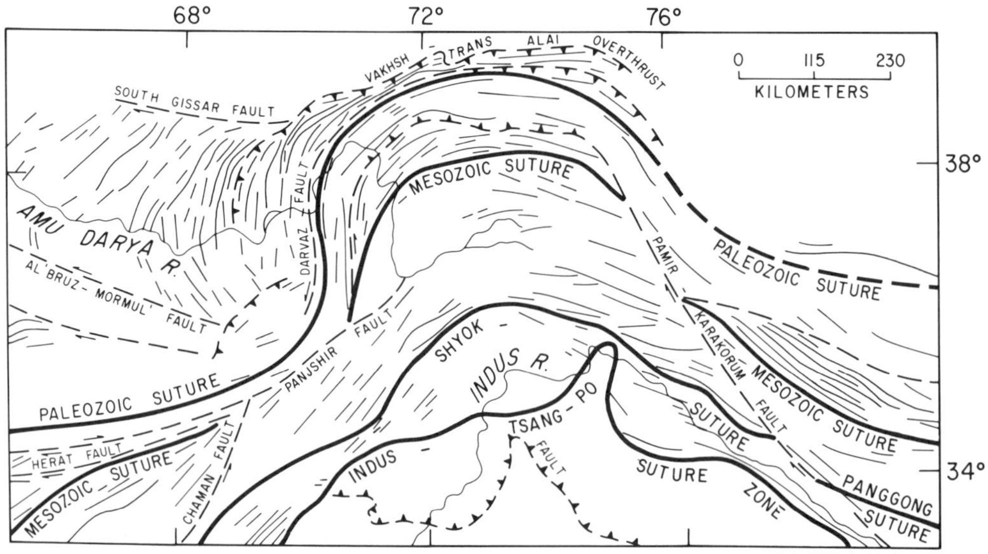

Figure 4. Structural map of the Pamir-Punjab syntaxis, Takjik Depression, and surrounding regions, based on the Geological Map of Afghanistan (1977), the Geological map of the Tadjik SSR and adjoining territories (1984), and the Geological map of the Xinjiang-Uyghur Autonomous Region, China (1985) and from data of Bratash and others (1970) and Desio (1965). Sutures of fragments of continental crust (ophiolite belts) are shown by heavy black lines. Axes of Cenozoic folds are shown by finer lines. Strike-slip faults are indicated by dashed lines. Major thrust faults are indicated by dark lines with teeth pointing in the down-dip directions.

1980; Boulin, 1981; Burtman, 1982; Montenat and others, 1986a; Ruzhentsev and Shvol'man, 1981a, 1981b; Sengör, 1979, 1984; Sengör and Hsü, 1984; Sengör and others, 1988; Stöcklin, 1977; Tapponnier and others, 1981). The histories of the various belts have been described by many geologists, and here we give only brief summaries of them. To measure the amount of penetration of the Pamir into the rest of Eurasia, we will concentrate on the Paleozoic suture.

The gross structure of the Pamir is commonly divided into four main zones: Northern, Central, Rushan-Pshart, and Southern Pamir. The Northern Pamir represents the late Paleozoic suture zone between the Central Pamir and the rest of Asia. The Rushan-Pshart zone marks the later suturing of the Southern Pamir to the Central Pamir (Fig. 3). Then in early Tertiary time, or possibly very late Cretaceous time, India collided with the southern margin of Eurasia, and the ocean basin between them, south of the Pamir, closed.

Late Paleozoic suturing of continental fragments to Asia. A late Paleozoic suture zone wraps around the Pamir, from the northern Hindu Kush in Afghanistan, through the northern Pamir, and then across the Kunlun of northern Tibet.

Northern Pamir. Early to middle Carboniferous (Tournaisian to Serpukhovian) igneous and sedimentary rock from an oceanic environment (ophiolites) crops out widely along the Northern Pamir (Fig. 3). In the eastern part of the Northern Pamir, in the area near the large lake Karakul, the base of the observed section consists of tholeiitic basalt about 1 km in thickness and tectonically overlying gabbro and ultrabasic rock (Pospelov, 1987). Small massifs of ultramafic rock, gabbro-diorite, and plagiogranite lie within the basalt (Budanov and Pashkov, 1988; Leven, 1981). In the upper part of the section, basalt alternates with sandstone, siltstone, and limestone, which contain foraminifera of Visean age (Budanov and Pashkov, 1988). All are overlain by argillites with bands of limestone containing brachiopods and foraminifera of Serpukhovian and Moskovian age (Leven, 1981). In the western part of the Northern Pamir, in the region of Kalaikhumb River in the Darvaz Range (Fig. 3), serpentine mélange is prevalent and overlain by a layer more than 2 km thick of pillow basalts of tholeiitic composition (Pospelov, 1987). In the upper part, the basalt contains limestone with goniatites of late Visean to early Serpukhovian age (Ruzhentsev and Shvol'man, 1981a; Ruzhentsev and others, 1977). The composition of the upper part of the Serpukhovian and Bashkirian layer varies in the Kalaikhumb region. One section shows basalt, a second contains island-arc volcanic rock (lavas and tuffs of andesitic-basalt, andesite, dacite, and liparite), and a third contains terrigenous sediment and olistostromes. A nappe comprising these rocks has been thrust northward onto the Eurasian margin and later compressed into folds that were thrust back southward, as retrocharriage (Pospelov and Sigachev, 1987). Conglomerate and limestone of Moscovian (late Carboniferous) age (Pospelov, 1987) and in places later Carboniferous and Permian deposits overlie this nappe. Thus, these igneous and sedimentary rocks record a pre–late Carboniferous ocean basin and its closure in late Carboniferous time.

Recent work on the eastern margin of the Pamir suggests that this zone continues southeastward. A zone of amphibolites, containing pillow basalts (Pan, 1992, p. 34), has been intruded by gneissic granite dated at 277 ± 6 Ma using U-Pb on zircons (Xu and others, 1992). Both have been overthrust onto Devonian and Carboniferous limestones and sandstones (P. Tapponnier and M. Brunel, personal communication, 1992). The age of overthrusting is almost surely younger than Paleozoic and could range from Jurassic to late Cenozoic (Brunel and others, 1992a).

Tectonic activity did not cease when the nappes in the Northern Pamir were emplaced. Packages of younger Carboniferous and Permian rock out in upright folds. The intrusion of Triassic-Jurassic granite into this deformed Permian rock, however, places a younger limit on the age of this phase of deformation, and possibly it had ceased before the Mesozoic Era began.

Untangling the detailed history of this Paleozoic belt is complicated by fragments of continental crust (microcontinental blocks) within the highly deformed oceanic rock of the Northern Pamir (Ruzhentsev and others, 1983). Andesitic rock, containing sediment with Tournaisian and Visean fossils, is found along the southern margin of the Northern Pamir and may represent remnants of an island arc or intracontinental rifting. An island-arc structure might also have been built on the continental slope of a fragment of continental crust that now defines the Northern Pamir (Fig. 3). This terrain, called the Kurgovat microcontinent (Ruzhentsev and Shyol'man, 1982), joined Eurasia as a result of the closing of the oceanic basin whose remnants consist of the basalt, gabbro, and untramafic rocks described above. A sharp difference between the foraminifera common to the Gondwanaland biogeographic province in the province in the Central Pamir and those of the Eurasian province found in the northern part of the Northern Pamir persisted until the Sakmarian stage of the early Permian period (Leven and Shcherbovich, 1978, pp. 63–82). The probable existence of a Carboniferous(?) to Permian ocean basin is further implied by mafic volcanic rock in the Akzhilga zone of the Northern Pamir (Fig. 3) (Budanov and Pashkov, 1988; Geological map of the Tadjik SSR and adjoining territories, 1984). Permian fusilinids are preserved in interbeds of sandstone within the volcanic rock (Leven, 1981). A part of the volcanic rock has been metamorphosed to albite-epidote-chlorite-actinolite schist (Belov and others, 1982), and numerous bodies of serpentinized peridotite, forming lenses up to 3 km long, gabbro, and plagiogranite lie within the volcanic rocks (Baratov, 1976, p. 219).

The Akzhilga zone appears to continue southeast into western China (Fig. 3). Metamorphosed basalt, including pillow basalt, gabbro, and pyroxenite crop out along the north-northwest–trending Tashkurgan Range, east and southeast of the town of Tashkurgan (Pan, 1992, p. 46). These rocks lie in close association with unfossiliferous metamorphosed black shale containing isolated blocks of granite. A glacial environment and a Permian age of the black shale seem likely.

The Central Pamir, which contains deformed and metamorphosed Precambrian and Paleozoic rock, constitutes another continental fragment that collided with Asia, probably in Permian time after the closure of the Akzhilga ocean basin.

Thus, the Northern Pamir seems to define a belt between the rest of Eurasia and the Central Pamir, where a major ocean basin closed and small continental fragments were accreted in Carboniferous to late Permian time.

Western Hindu Kush. The Paleozoic geologic history of the Hindu Kush west and north of Kabul resembles that of the Northern Pamir. The suture zone can be traced from the Northern Pamir through northern Afghanistan and to the northern edge of the western Hindu Kush. In the basin of the Surkhab River (Fig. 3), early Carboniferous volcanic rock, roughly 2.5 km thick, contain pillow basalt in the lower part of the section, and basalt, andesite, volcanic breccia, and tuff alternate higher in it. This section lies on limestone containing brachiopods, corals, and foraminifera of Visean age and is unconformably covered by conglomerate and limestone with Bashkirian foraminifera (Pyzh'yanov and Sonin, 1977; Pyzh'yanov and others, 1978). Thin bodies of ultrabasic rock also occur within the early Carboniferous deposits in this region (Kolchanov and others, 1971). Scattered outcrops of serpentinite, which presumably originated as ultramafic parts of ophiolite suites, crop out along the Herat fault (Boulin and Bouyx, 1977; Stöcklin, 1977), an active right-lateral strike-slip fault through the western Hindu Kush (e.g., Wellman, 1966). These outcrops of serpentinite probably are fragments of material later thrust southward from the suture farther north. Much of the rock exposed in the Hindu Kush north and west of Kabul is early to middle Paleozoic metamorphic rock. Deformation seems to have occurred in early Carboniferous time, apparently during its suturing to Eurasia (Boulin, 1981; Boulin and Bouyx, 1977; Stöcklin, 1977). As in the Northern Pamir, late Carboniferous, Permian, and younger deposits unconformably overlie both older metamorphic rock (e.g., Boulin and Bouyx, 1977) and ophiolites that mark the southwestward continuation from the Northern Pamir (Pyzh'yanov and others, 1978). Thus, the history of tectonic activity is similar to that in the Northern Pamir: An ocean basin closed during the Carboniferous Period, and the sutured margins of the continents underwent pronounced crustal shortening.

Kunlun. Late Paleozoic ophiolites also have been found in the Kunlun along the northern margin of western Tibet (Bian, 1992; Matte and others, 1992; Norin, 1946). Norin (1946, p. 16) wrote that within the Devonian sequence is "a great complex of various porphyrites, keratophyres, tuffaceous beds, and agglomerates," which he compared with the Franciscan complex in California. Sinitsyn (1959, p. 414) also assigned a Devonian age to a section of volcanic rock found in the Khotan River basin in the western Kunlun; the lower part contains spilite and radiolarite, the upper part lavas and tuff of intermediate to acidic composition. Norin described a Carboniferous series unconformably overlying the folded Devonian rock and consisting of conglomerate grading upward into arkosic sandstone, limestone, and shale. These observations suggest an eastward continuation of the Paleozoic suture.

The explanatory notes accompanying the Geological map of the Xinjiang-Uygur Autonomous Region, China (1985, p. 80) assign radiometric ages (using unspecified isotopes) of 274 to 298 Ma to granodiorite intrusions into late Paleozoic ("Variscan") mafic and ultramafic bodies in the Kunlun and farther east in the Altyn Tagh (Geological map of the Tibetan Plateau, 1980; Geological map of the Xinjiang-Uygur Autonomous Region, China, 1985; Sinitsyn, 1959). These outcrops presumably mark ophiolitic remnants of the same ocean basin traced farther west in the Pamir and the Hindu Kush. This zone may continue eastward (e.g., Sengör, 1984) into areas not yet comprehensively mapped.

Late Mesozoic suturing of continental fragments to Asia. Traces of Mesozoic suturing can be found along the Rushan-Pshart zone and in the Central and Southern Pamir. Although continuity is interrupted, evidence of Mesozoic suturing also can be found in western Afghanistan and in Tibet.

Rushan-Pshart zone and the Southern and Central Pamir. The sequence of rock in the Central Pamir (Fig. 3) reveals a Paleozoic history culminating with unconformable deposition of late Carboniferous and early Permian sandstone, limestone, and marl. The continued deposition of detrital and carbonate sediments in shallow water in the Central Pamir, from late Paleozoic to Jurassic time, suggests the presence of a platform or passive continental margin. Evidence of volcanism apparently has not been found. Moreover, in most areas Cretaceous and Paleogene sedimentary rock was deposited without marked angular discordance. In a few places this angle reaches 10 to 15°, but in general it is too small to be measured. In some areas, the sequence is nearly complete from late Carboniferous to Paleogene, but in others late Jurassic and early Cretaceous rock is missing (Ruzhentsev, 1971, p. 13). Ruzhentsev (1971, p. 79) did describe a local unconformity of late Cretaceous limestone on intensely folded Triassic and Jurassic sandstone and slate from two areas, the Muzkol and Yazgulem ranges, but evidence for the formation of large overthrusts between late Carboniferous and Paleogene time is absent. As discussed below, the major overthrust faults in the Central Pamir formed in Cenozoic time.

The Rushan-Pshart zone, between the Central and Southern Pamir, marks a Mesozoic ocean basin within the Pamir (e.g., Pashkov and Shvol'man, 1979; Shvol'man, 1978). Remnants of the northern margin of this basin are distributed throughout the Rushan-Pshart zone. Limestone characterizes the basal part of the late Permian and Triassic deposits, but toward the top the section includes predominantly pillow basalt, andesite-basalt, and tuff. Small bodies of ultramafic rock are associated with basalt. The volcanic rock alternates with radiolarite and limestone. Deep-water terrigenous deposits of Jurassic and Cretaceous age overlie 1.5 km of Triassic volcanic rock (Pashkov and Shvol'man, 1979). Shvol'man (1978; Pashkov and Shvol'man, 1979) emphasized clear variations in rock types and facies along strike but showed that the whole of the belt marks an important zone of latest Paleozoic to early Mesozoic rifting and the formation of a basin with some characteristics of an ocean basin.

This basin closed in the late Jurassic or early Cretaceous period, when a few thrust sheets were emplaced *within* the Rushan-Pshart zone. The thrust sheets themselves were only slightly deformed. Red Cretaceous sediment was deposited on the rock within thrust sheets with only slight angular unconformities (<15°).

In the North Alichur Range (Fig. 3), metamorphic rock of the Southwestern Pamir underlies remnants of a nappe of basalt, picrite, and their tuffs, which contain limestone with corals and ammonites of Ladinian and Carnian stages (Triassic) (Dronov, 1988). These fauna have been assigned a "South Alpine" type. The thrust sheet appears to be rooted in the Rushan-Pshart zone. According to Pashkov and Budanov (1990), both the thrust sheet and the metamorphic rock are overlain by early Jurassic (Hettangian) deposits. Thus, the nappe, which consists of a slice of oceanic crust, was emplaced in late Triassic time.

Significantly thicker Mesozoic ophiolites are found in the South Alichur Range (Fig. 3) near the boundary between the Southeastern and Southwestern Pamir. They too lie in tectonic contact with the metamorphic rock. Hettangian-Sinemurian deposits overlie these ophiolites with an angular unconformity (Dronov, 1986). Shvol'man (1980) noted that the lower part of the ophiolite consists of serpentinized harzburgite, cut by dikes of gabbro, gabbro-diorite, and plagiogranite. Overlying the ultramafic tectonite is a sequence of alkali-olivine basalt and tholeiitic basalt with layers of siliceous shale and tuff containing Mesozoic phytoplankton. (A more precise age cannot be given.) This sequence has a thickness of more than 1 km. Higher in the section are acidic lava, tuff, and olistostrome with blocks containing Permian and Triassic fauna up to the Carnian stage. The ophiolite consists of tectonic slices. According to Shvol'man (1980), these thrust sheets formed simultaneously with the deposition of the olistostromes.

To the south of the Rushan-Pshart zone, the Southern Pamir is necessarily divided into Southeastern and Southwestern regions because of the completely different rocks exposed. The Southwestern Pamir consists of Precambrian metamorphic rock and Mesozoic and Paleogene granite (Pashkov and Budanov, 1990). The oldest rock in the Southeastern Pamir is late Carboniferous to early Permian sandstone, siltstone, clay, and limestone. The Triassic sequence of limestone, radiolarite, and siltstone contains rare basaltic lava and tuff, like those in the Rushan-Pshart zone but less abundant. Jurassic reef limestone unconformably overlies the Triassic rock and is overlain, in turn, by Cretaceous deposits that include conglomerate as well as dacite, andesite, tuff, and limestone. Folding in the Southeastern Pamir and Rushan-Pshart zone occurred in Jurassic and probably early Cretaceous time, but all tectonic structures were strongly reactivated in Cenozoic time, when deformation occurred throughout the Pamir.

The sequences in the Southern and Central Pamir are quite different. The Rushan-Pshart zone seems to mark the suturing of the Southern to the Central Pamir but possibly not the closing of a large ocean.

Mesozoic suture in Afghanistan. Neither the stratigraphic sequences of the Southwestern Pamir nor the ophiolitic complex of the Rushan-Pshart zone are represented in northeastern Afghanistan (Desio, 1975), but the Rushan-Pshart zone seems to correlate with the Farah Rud basin in Afghanistan (Boulin, 1981; Burtman, 1982; Girardeau and others, 1989; Karapetov and others, 1975; Montenat and others, 1986a; Ruzhentsev and Shvol'man, 1981a; Sengör, 1984; Shvol'man and Pashkov, 1986; Tapponnier and others, 1981).

The Khash Rud (Waras or Panjaw) zone along the southwest boundary of the Farah Rud basin (Fig. 3), in west-central Afghanistan, contains turbidites with volcanic rock at various levels. Dronov (1980, pp. 182–183) stated that two formations, the Alekon and the Navzod, characterize this region, but the relation between them is not clear. The Alekon Formation, up to 4 km thick, consists of sandstone and siltstone with horizons of limestone, andesite, conglomerate, and basalt. Plant remains of Malm age have been discovered in the volcanogenic terrigenous deposits of the Khash Rud (Waras) zone, and middle or late Triassic foraminifera in the lower parts of the section (Blaise and others, 1978; Girardeau and others, 1989). Late Triassic fossil fauna have also been found in olistoliths from the lower part of this section (Girardeau and others, 1989). A sequence 500 m thick of red conglomerate, sandstone, and limestone with orbitolina and coral of Aptian and Albian age overlies this with angular unconformity (Blaise and others, 1978; Dronov, 1980, p. 195). Thus, the ages of this volcanogenic sequence span the late Triassic to Aptian. The upper member of the Mesozoic section consists of 700 m of red terrigenous continental sediment that forms an erosional unconformity on early Cretaceous rock and probably is late Cretaceous in age (Dronov, 1980, p. 209). The Navzod Formation, about 1.5 km thick, is composed largely of andesite and basalt (in part, pillow basalt), but layers of chert, sandstone, and limestone contain gastropods with an age interval of Tithonian to Cenomanian. The volcanic rock is also associated with ultramafic rock, gabbro, and diorite. The ultramafic rock forms bodies up to 1 km wide and 18 km long and consists of serpentinite, serpentinized lherzolite, spinel peridotite, and dunite. Gabbro, gabbro-norite, gabbro-diabase, and diorite form sills, stocks, and dikes within the volcanic sequence (Dronov, 1980, pp. 287–290; Girardeau and others, 1989). Thus, the Khash Rud ophiolite zone probably marks the final suturing across a Mesozoic ocean and presumably is equivalent to the Rushan-Pshart zone.

Sediment cropping out to the north and west, over a large part of the Farah Rud basin, apparently was deposited on a passive continental margin of this ocean basin. The oldest deposits contain Carboniferous and Permian fossil fauna. This 5-km-thick sequence consists of polymictic sandstone, argillaceous slate, and siltstone. Poorly preserved fusilinids occur in bands of limestone in the upper part (Dronov, 1980, p. 110). Late Permian foraminifera are also present in the limestone, which crops out in isolated remnants of nappes some 150 m thick (Dronov, 1980, p. 131). The Triassic sequence consists of 2 km of alternating sandstone, siltstone, argillaceous slate, conglomerate, chert, andesite, basalt, tuff, and limestone with Carnian and Norian pelecipods and corals (Dronov, 1980, p. 141). Higher in the stratigraphic sequence are 2 km of polymictic sandstone, argillaceous slate, and siltstone, over which 150 m of Dogger limestone lie unconformably (Dronov, 1980, pp. 155, 172). Like the southern edge of the Central Pamir, this sequence could mark a passive continental margin.

Flysch, reaching 7 to 9 km in thickness, occupies a large part of the Farah Rud basin and implies the presence of a deep basin. In the middle and upper parts, the sequence contains ammonites, pelecipods, brachiopods, and foraminifera of Berriasian, Valanginian, and Hauterivian age (Dronov, 1980, pp. 186–187). Barremian and Aptian rock is presented by organic limestone, about 1 km thick. On it, separated by a disconformity, lie red conglomerate, sandstone, and limestone with Aptian and Albian fossils (Dronov, 1980, pp. 193–295). Above another erosional unconformity lie red conglomerate, sandstone, and siltstone barren of fossils but surely of late Cretaceous age (Dronov, 1980, p. 209). Then, Paleogene andesite, dacite, tuff, and sandstone lie on Cretaceous rock with sharp angular discordance. Thus, like the Rushan-Pshart zone, the Farah Rud basin apparently began to form in latest Paleozoic and closed in mid-Cretaceous time. It is not obvious that the Farah Rud basin and the Khash Rud ophiolite are the remnants of a wide expanse of ocean floor.

The Central Mountains of Afghanistan, southeast of the Farah Rud basin, lie in thrust contact on the southern edge of the eastern part of that basin. The eastern part of the Central Mountains reveals a virtually complete Paleozoic series of platform sedimentation. In particular, late Carboniferous and Permian tillite, coal, and "coal fauna" indicate that this area was part of Gondwanaland (Blaise and others, 1977). None of this record is present in the Southern Pamir, except possibly in its southeastern extremity. A thick series of Permian to Jurassic calcareous rock, however, overlies these older rocks in Afghanistan and grossly resembles a similar sequence in the Southeastern Pamir. Moreover, the Permo-Triassic series is absent in some parts of the Central Mountains of Afghanistan, as it is in the Southwestern Pamir. Blaise and others (1977) emphasized that no important disruption, or angular unconformity, interrupts this sequence of Paleozoic and Mesozoic rock; only at the end of the Jurassic Period did an interruption occur in the Central Mountains of Afghanistan.

The Farah Rud basin is quite wide (200 km) in western Afghanistan, which raises the possibility that it never closed completely. Farther east, however, it is overridden by the Central Mountains (Fig. 3). Its presumed continuity with the Rushan-Pshart zone apparently has been disrupted by strike-slip displacements on the Chaman, Herat, Panjshir (or Zebak-Munjan), and other faults (Fig. 4) (Dronov, 1980, pp. 464–466; Tapponnier and others, 1981).

Central Tibet. Possible eastward continuations of the Mesozoic basin revealed by the Farah Rud and Rushan-Pshart zones project across the central Tibetan Plateau (Burtman, 1982; Girardeau and others, 1984; Peive and others, 1964; Sengör, 1984; Shvol'man and Pashkov, 1986). Belts of ophiolites are commonly described as extending eastward both from just south of the Kunlun (e.g., Chang and Cheng, 1973; Pearce and Deng, 1988; Sengör, 1984; Stöcklin, 1977) to northeastern Tibet and from the lake, Panggong (or Banggong), in western Tibet (e.g., Chang and Pan, 1981; Matte and others, 1992) to the Dongqiao ophiolites some 1,500 km farther east (e.g., Chang and Cheng, 1973; Girardeau and others, 1984; Pearce and Deng, 1988). We tentatively associate the northern of these, herein called the west Kunlun suture, with the Rushan-Pshart zone and the Panggong suture with the Shyok suture (Fig. 4), but this is not the common association. Among all of the pre–late Cretaceous ophiolite suites in Tibet, the most thoroughly studied seems to be the Dongqiao suite, where a late Jurassic to early Cretaceous age of emplacement of the ophiolites seems to be well established. The nearly total absence of plutonic activity associated with it (e.g., Harris and others, 1988a, 1988b; Xu and others, 1985) suggests that subduction may not have been long-lived or deep. If the Dongqiao ophiolite suite is properly correlated with the Banggong ophiolite, as is commonly inferred, then our suggested correlations are incorrect. From the brief descriptions of the Panggong (Chang and Pan, 1981; Matte and others, 1992) and the west Kunlun (Pan, 1992) sutures, however, correlations of either with Dongqiao seems open.

Unlike the Paleozoic suture in the Northern Pamir and its continuations east and west, the Rushan-Pshart zone and its continuations into Afghanistan and Tibet seem to mark a zone of localized convergence and ophiolite emplacement. A major oceanic terrain may not have been consumed. Although ophiolites and other remnants of ocean floor do not crop out continuously along this zone, a relatively continuous belt seems to be present and deflected around the western syntaxis of the Himalaya.

Indus–Tsang-po suture zone. A third belt of ophiolites follows the Tsang-po and Indus valleys in southern Tibet, wraps around the northwestern margin of the Indian subcontinent, and seems to continue into southern Pakistan, Afghanistan, and Iran (e.g., Gansser, 1964, 1979, 1980; Montenat and others, 1986b; Searle, 1991; Stöcklin, 1977; Tahirkheli and others, 1979; Tapponnier and others, 1981). In the west, this belt diverges into separate ophiolitic belts (e.g., Shyok suture and its continuations) that surround fragments of continental crust or ancient island arcs (e.g., Andrews-Speed and Brookfield, 1982; Searle, 1991; Thakur and Misra, 1984) (Fig. 4). Together, the Indus–Tsang-po and Shyok ophiolite belts seem to mark the suture between the Indian subcontinent with southern Eurasia and islands arcs in early Cenozoic (or possibly latest Mesozoic) time. The Himalaya, some 200 to 300 km in width, consist largely of rock scraped off of the ancient northern margin of India as the Indian subcontinent plunged northward beneath southern Tibet. Thus, the Indus–Tsang-po suture zone probably resembles the precollisional shape of India's margin more than that of southern Eurasia. Accordingly, the deflection of the ophiolite belts around the northwestern prong of India probably represents a penetration of that prong into what originally was a straighter southern margin of Eurasia.

Summary of deflections of sutures around the Pamir. Remnants of three ancient ocean basins can be traced across Afghanistan, northward around the Pamir and the Punjab syntaxis of the Himalaya, then south again and across the Tibetan Plateau. Two of them, the northern late Paleozoic suture and the early Cenozoic (or perhaps locally, latest Mesozoic) Indus–Tsang-po suture, seem to mark zones where wide expanses of oceanic lithosphere were subducted and where fragments of continental crust traveled long distances to collide with southern Eurasia. The third, represented by the Farah Rud basin in Afghanistan, the Rushan-Pshart zone in the Pamir, and apparently the Dongqiao ophiolites in Tibet, might mark narrow, even disconnected, basins and not necessarily the suturing of widely separated continental fragments. In any case, whereas all three belts trend roughly east-west and are spaced far apart both in Iran and western Afghanistan and in Tibet, they are markedly deflected around the Punjab syntaxis of the Himalaya, and north of it they lie closer together than farther east or west. Thus, these segments north of the syntaxis and in the Pamir seem to have been displaced northward with respect to their extensions east and west. Clearly, the Punjab-Pamir syntaxis arose as a result of the collision of India with Eurasia.

Not only does the Indus–Tsang-po suture zone bend around the northern prong of the Indian subcontinent, but so do the axes of Cenozoic folds. Together they outline two arcs. The Hindu Kush–Karakorum arc lies between the Cenozoic and Mesozoic sutures (Fig. 4). The Pamir, or external, arc is outlined by the structures of the Pamir, the western Kunlun, and the area northeast of the Hindu Kush and approximately embraces the area between the Paleozoic and Mesozoic sutures.

The external arc is more compressed than the internal arc and is asymmetric, for its western side is the more sharply bent. On the flanks of the Pamir arc, Cenozoic fold axes outline two broad shear zones, or sigmoidal structures in plan view: the Afghan-Pamir left-stepping sigmoid on the west in northern Afghanistan to the Northern Pamir and the Pamir-Kunlun right-stepping sigmoid on the east across the Northern Pamir to the western Kunlun. The limbs of each are more than 300 km long (Figs. 3 and 4) and define the magnitude of northward displacement involved in the formation of the Pamir arc.

Paleomagnetic declinations and the bending of the Pamir arc

Paleomagnetic measurements from sites in the outer zone of the Pamir corroborate the inference of bending of the structural belts (Bazhenov and Burtman, 1981, 1982, 1986, 1990; Pozzi and Feinberg, 1991). Declinations in samples from early Cretaceous and Eocene-Oligocene sedimentary rocks along the outer edge of the Pamir show a smooth variation from a northwesterly orientation in the western Pamir, to roughly due north in the northwestern Pamir, to due east at the northern edge of the range (Fig. 5; Tables 1 and 2). Measured declinations of early Cretaceous and Paleogene rocks from the Kopet Dagh, some 1,000 km to the west (Fig. 6), from the South Tien Shan (Fig. 5), and from early and late Cretaceous sedimentary rocks of the Tarim Basin (Chen and others, 1993) lie slightly east of north and point toward the early and late Cretaceous and Paleogene poles for Eurasia. Thus, the declinations from the Pamir outer arc record significant rotations about vertical axes—counterclockwise in the west and clockwise in the north—with respect to Eurasia.

The Cretaceous and Tertiary sequences that were sampled appear to overlie, in stratigraphic contact, Paleozoic rock

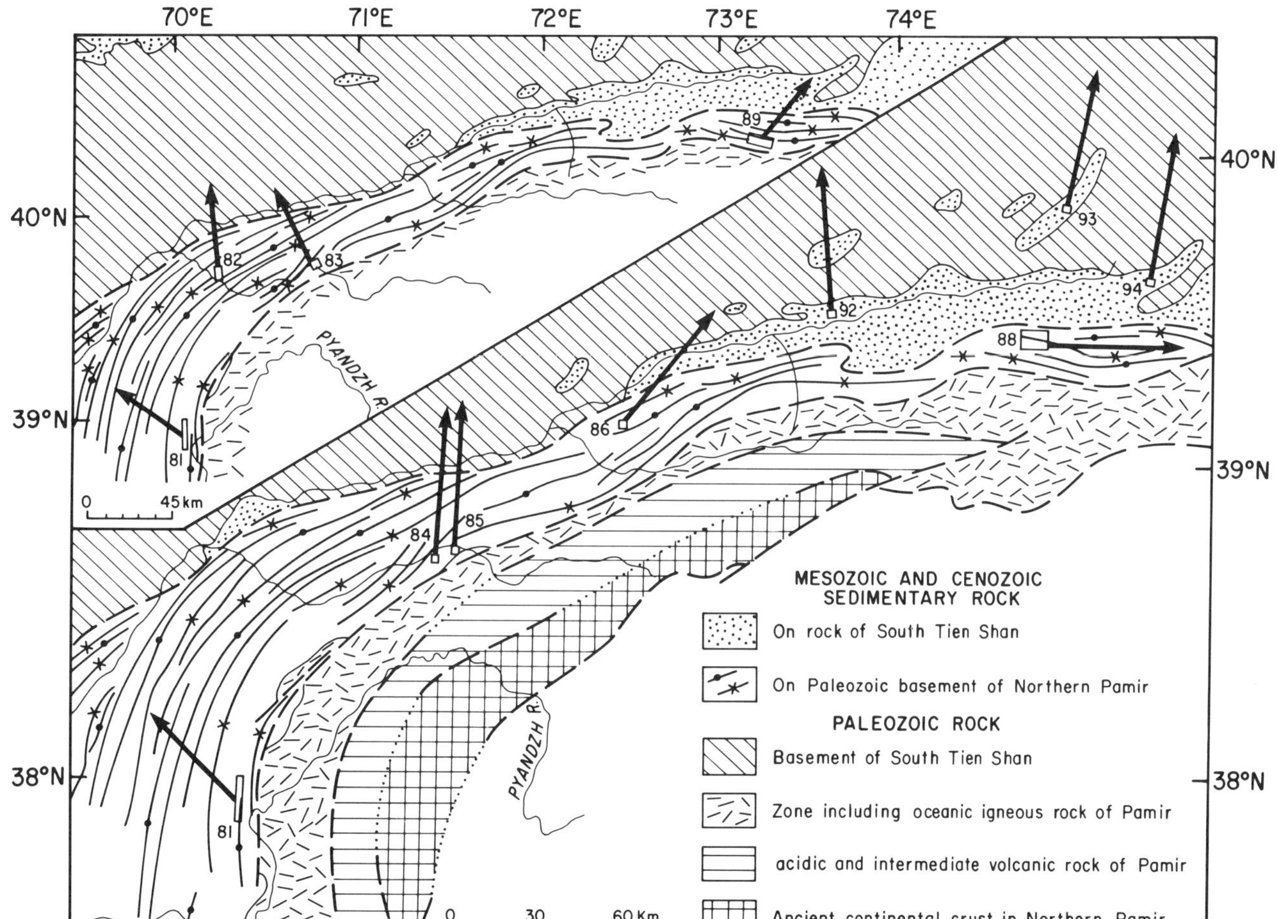

Figure 5. Map showing paleomagnetic declinations measured in Paleogene to early Miocene (upper) and in early Cretaceous (lower) sedimentary rock in the Tadjik Depression and Pamir. (Redrawn from Bazhenov and Burtman [1990, p. 121]) Note how magnetic declinations, indicated by arrows, from sites of the outer Pamir zone in the west point northwest and those in the northeast point east. Magnetizations from sites north of the outer zone of the Pamir (no's. 92, 93, and 94) point north toward the early Cretaceous pole of Eurasia. Numbers next to arrows are the same as in Figure 7. Lines in the areas of Mesozoic and Cenozoic sedimentary rock in the external zone of the Pamir show fold axes of anticlines (with dots) and synclines (with crosses).

TABLE 1. PALEOMAGNETIC RESULTS FOR CRETACEOUS ROCKS IN THE PAMIR AND ALAI REGIONS*

Sample Locality†	H§ (m)	n_o§	n§	Declination (°)	Inclination (°)	k§	α95§ (°)
81a	250	16	11	310	46	50	6.0
81b	180	14	12	317	41	55	5.5
81c	400	15	9	317	44	61	6.0
Mean (81)	......	......	32	314	44	54	3.4
81**	40	13	5	309	37	65	7.8
84a	120	19	14	3	43	29	7.0
84b	300	38	32	11	45	72	2.9
Mean (84)	......	......	46	8	44	48	3.0
85a	200	34	34	3	54	104	2.4
85b	30	10	6	202	-68	4	16.7
86	200	32	30	40	30	......	12
88	......	......	(5)	92	51	......	10
88, 90, 91	......	......	(7)	89	51	......	10
92	100	15	10	359	51	27	8.5
93a	100	7	6	12	47	29	10.7
93b	500	22	17	8	52	30	6.2
Mean (93)	......	......	23	9	50	30	5.3
94	250	22	19	14	45	25	6.5
Mean (92-94)	......	......	42	8	49	226	4.7

*Bazhenov and Burtman, 1990.
†Sample localities are shown in Figure 5.
§H = thickness of the sedimentary sequence; n_o = number of samples measured; n = number of samples used; where surrounded by parentheses, gives the number of demagnetization paths that define a secondary component; k = precision parameter; α95 = radius of the 95 percent confidence region.
**Late Cretaceous samples.

TABLE 2. PALEOMAGNETIC RESULTS FOR PALEOGENE ROCKS IN THE OUTER ZONE OF THE PAMIR*

Sample Locality†	H§ (m)	n_o§	n§	Declination (°)	Inclination (°)	k§	α95§ (°)
81	200	54	39	305	29	11	6.8
82a	120	24	18	354	30	19	7.6
82b	100	15	14	325	52	16	9.4
82c	80	14	14	349	30	7	14.2
Mean (82a + 82b)	......	......	32	352	30	12	7.2
83	200	39	31	329	40	12	7.3
89a	100	12	9	39	32	19	10.6
89b	150	12	11	45	43	12	12.3
89c	50	12	7	30	50	9	17.8
89d	70	31	27	35	39	13	5.3

*Bazhenov and Burtman, 1990.
†Sample localities are shown in Figure 5.
§H = thickness of the sedimentary sequence; n_o = number of samples measured; n = number of samples used; where surrounded by parentheses, gives the number of demagnetization paths that define a secondary component; k = precision parameter; α95 = radius of the 95 percent confidence region.

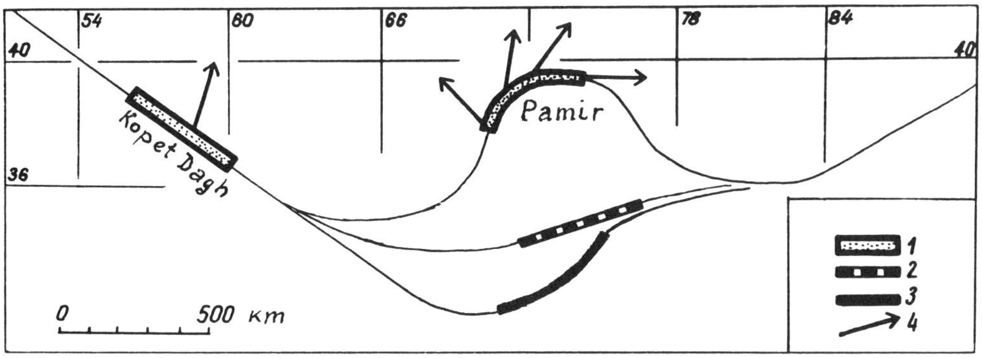

Figure 6. Map showing present and reconstructed positions of the outer Pamir arc from paleomagnetic data. (From Bazhenov and Burtman [1990, p. 137].) (1) Present position of the outer arc of the Pamir. (2) Reconstructed position for the beginning of Neogene time. (3) Reconstructed position for early Cretaceous time. (4) Summary of paleomagnetic declinations from early Cretaceous sites.

of the Northern Pamir. These Cretaceous and Tertiary rocks were deposited on the southern margin of the Tadjik sedimentary basin, whose deposits now lie within the Tadjik Depression and the Trans-Alai Range.

The strata sampled by Bazhenov and Burtman crop out in folds whose axes trend parallel to the edge of the Pamir. The declinations vary smoothly with the trends of fold axes (Fig. 7). Bazhenov and Burtman (1981, 1982) assumed that folding preceded the rotation of declinations. If the Cretaceous and Tertiary sequences along the outer edge of the Pamir and across northern Afghanistan to the Kopet Dagh defined a line of constant length that has subsequently been bent into its present shape, then the declination anomalies from the Pamir would constrain how much that line has been bent. Calculated families of early Cretaceous and Paleogene positions of the outer margin of the Pamir suggested that the Pamir arc would have lain between 300 and 700 km south of its present limit (Fig. 7), for which declinations from the Kopet Dagh define a reference frame fixed to Eurasia. Thus, these reconstructed positions concur with the inference that the late Paleozoic and late Mesozoic sutures have been displaced 300 km or more northward with respect to their continuations in Afghanistan and Tibet. Unfortunately, the scatter of inclinations and errors in them, probably due to compaction of the sediment, do not allow a further test of this displacement (Bazhenov and Burtman, 1990).

Independently, Pozzi and Feinberg (1991) studied late Cretaceous red beds from four sites within the Tadjik Depression. Their results were like those of Bazhenov and Burtman (1990): Samples from the site east of the Vakhsh Overthrust show a large counterclockwise rotation of 52 ± 10°. Similarly, those from the three sites east of the Vakhsh Overthrust indicate counterclockwise rotation of only 20 ± 22°, 10 ± 9°, and 17 ± 9°. Pozzi and Feinberg (1991) averaged results from all four sites to obtain a virtual geomagnetic pole, but given the resolvably different amounts of rotation, we consider this un-

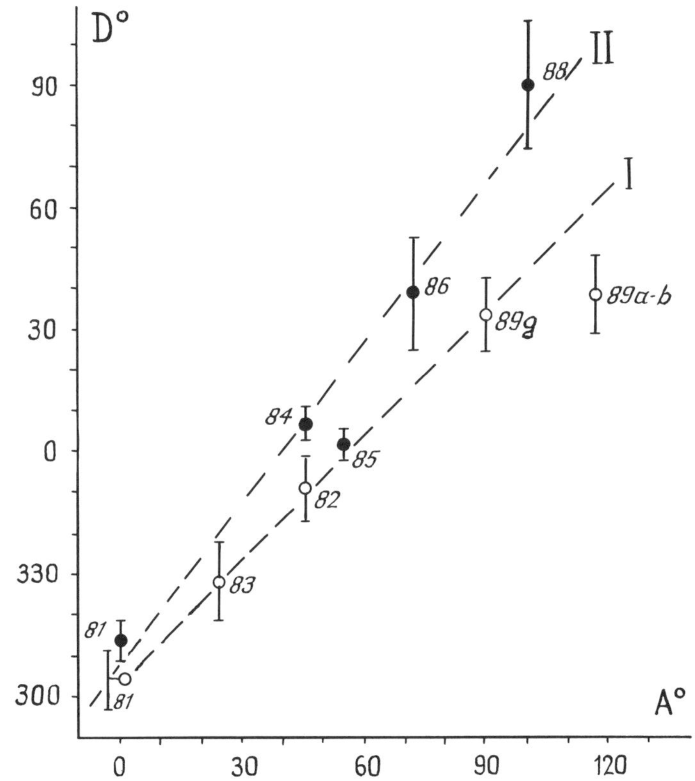

Figure 7. Plot of measured paleomagnetic declinations (D) versus trends of fold axes (A) from the outer zone of the Pamir. (From Bazhenov and Burtman [1990, p. 130].) Closed symbols are from early Cretaceous samples: open symbols are from Eocene, Oligocene, and early Miocene samples. Numbers refer to the localities shown on Figure 5. The slope of unity for the dashed line (I) passing through most of the Paleogene–early Miocene data implies that the fold axes have been rotated by the amount given by the declination anomaly and that the folds were linear and roughly parallel before early Miocene time. The steeper slope of the dashed line (II) through the data for early Cretaceous samples suggests that some rotation of the rock in the folds occurred before Paleogene time and that the greater the post-Cretaceous rotation the greater the pre-Paleogene rotation.

wise. The measured early Cretaceous paleolatitudes lie well within those appropriate for Eurasia, suggesting that post-Cretaceous convergence with Eurasia has been minor.

Cenozoic crustal shortening within the Pamir

In its northward advance, the Pamir has not behaved as a rigid body. Thrust faulting and folding of late Cretaceous and Paleogene rock *within* the Pamir show that the overthrusting of the Pamir onto the ancient eastern part of the Tadjik Depression has not been the only process by which the Pamir and the South Tien Shan have converged. Nappes, thrust sheets, and folds in the Southern Pamir, the Rushan-Pshart zone, and the Central Pamir provide quantitative evidence for a significant amount of Cenozoic crustal shortening. The Northern Pamir also was subjected to Cenozoic shortening, but quantifying the amount in the Paleozoic rock of this zone is difficult. The estimates of shortening are based largely on the measured displacements of thrust sheets, or nappes, and on analyses of folded structures. In addition, many of the major thrust faults truncate strike-slip faults, for which amounts of displacement can be determined accurately from offsets of tectonic structures, facies boundaries, and other geologic features. These strike-slip offsets allow refinements of estimates of crustal shortening based solely on minimum displacements of overthrust sheets. (Recall that offsets on strike-slip faults provide some of the tightest constraints on magnitudes of extension in the Basin and Range Province [Davis and Burchfiel, 1973; Wernicke and others, 1982].)

Southern Pamir and Rushan-Pshart zone. Pashkov and Budanov (1990) reported a large displacement of a thrust sheet, or nappe, onto the Southeastern Pamir. The footwall, exposed throughout the Southwestern Pamir, consists of Precambrian and Paleozoic metamorphic rock on which remnants of a Paleogene(?) sedimentary cover are preserved. (To avoid any suggestion that some patch of ground within the Pamir defines a geodynamically meaningful fixed reference frame, we use the term *footwall* where in older literature *autochthon* might have been used.) The metamorphic rock is intruded by granite and diorite assigned Triassic, Cretaceous, and Paleogene ages. The thrust sheet, exposed throughout much of the Southeastern Pamir and in klippen overlying the metamorphic rock of the Southwestern Pamir (Fig. 8), consists of a Carboniferous to Triassic sequence unconformably overlain by Jurassic, Cretaceous, and Paleogene sedimentary rock. According to Pashkov and Budanov (1990), this thrust sheet covers Cretaceous and Oligocene granites intruded into the metamorphic rock. The distribution of outcropping remnants of the sheet suggests that prior to erosion, it covered a large part, if not all, of the territory defined as the Southwestern Pamir. The north-south dimension of the exposed thrust sheet of 110 km gives a minimum displacement on the underlying thrust fault (Fig. 8). Pashkov and Budanov (1990) suggested that the overthrusting of this rock could have begun in the Mesozoic era and then been renewed in Miocene time. Evaluating the contribution of Mesozoic displacement of the Southeastern Pamir thrust sheet, if any contribution exists, does not, however, appear possible.

Other thrust faults are found within the Southeastern Pamir, where they divide the Carboniferous to Jurassic sedimentary sequence into smaller thrust sheets within the major one. Ruzhentsev (1968, pp. 107–114) measured displacements of 15 to 20 km on each of the largest thrust faults, the north-verging Murgab fault (Figs. 8 and 9) and the south-verging Gurumdi fault (Fig. 8). He estimated the total displacement of the smaller thrust sheets *within* the Southeastern Pamir to be 50 km or more (Ruzhentsev, 1990b). Moreover, the smaller thrust sheets also were shortened by folding after their emplacement (Fig. 10). Thus, the north-south convergence across the Southern Pamir, due both to the emplacement of the large Southeastern Pamir thrust sheet over the footwall exposed in Southwestern Pamir and to thrust faulting and folding within this sheet, apparently exceeds 160 km.

In addition, a system of northwest-trending right-lateral strike-slip faults, the Aksu-Murgab fault system, cuts the thrust sheets within the Southeastern Pamir nappe. Strike-slip displacements can be reliably determined from the offset of facies zones in the Carboniferous, Permian, Triassic, and Jurassic sedimentary rocks (Ruzhentsev, 1968, pp. 119–152). Both Jurassic facies zones and older thrust faults have been displaced right-laterally (Ruzhentsev, 1963, 1968, pp. 119–129): about 20 km on the Istyk fault and about 60 km in total on the Karasui and Lower Murgab faults (Fig. 8). At this northern end, the Aksu-Murgab system of right-lateral faults is truncated by the Lyangar-Sarez thrust fault (Fig. 8), along which rock of the Southeastern Pamir was thrust beneath the Rushan-Pshart zone and the Central Pamir. The demonstrable offset on the Lyangar-Sarez thrust is only 20 km (Peive and others, 1964). The total north-south underthrusting, however, is probably closer to 50 km, equal to the north-south component of the 80-km cumulative strike slip along the northwest-trending Aksu-Murgab fault zone. This estimate assumes no major rotation about vertical axes of the strike-slip faults with respect to the thrust faults.

Another strike-slip fault, the East Pamir fault (Fig. 8), cuts the Southeastern Pamir and the Rushan-Pshart zone and displaces them 30 km from their eastward continuations (Ruzhentsev, 1968, p. 145). This strike-slip displacement is absorbed in part by thrust slip along the Pshart fault (Fig. 8), where the Rushan-Pshart zone is thrust northward onto Tertiary deposits of the Central Pamir (Fig. 11). Thus, the cumulative Cenozoic shortening by thrust faulting along the margins of the Rushan-Pshart zone reaches 80 km. Convergence across the Southern Pamir and Rushan-Pshart zone, largely if not entirely Cenozoic in age, appears to be 240 km, or more.

The Aksu-Murgab and East Pamir faults constitute splays of the northwest-trending Pamir-Karakorum fault (Figs. 3 and

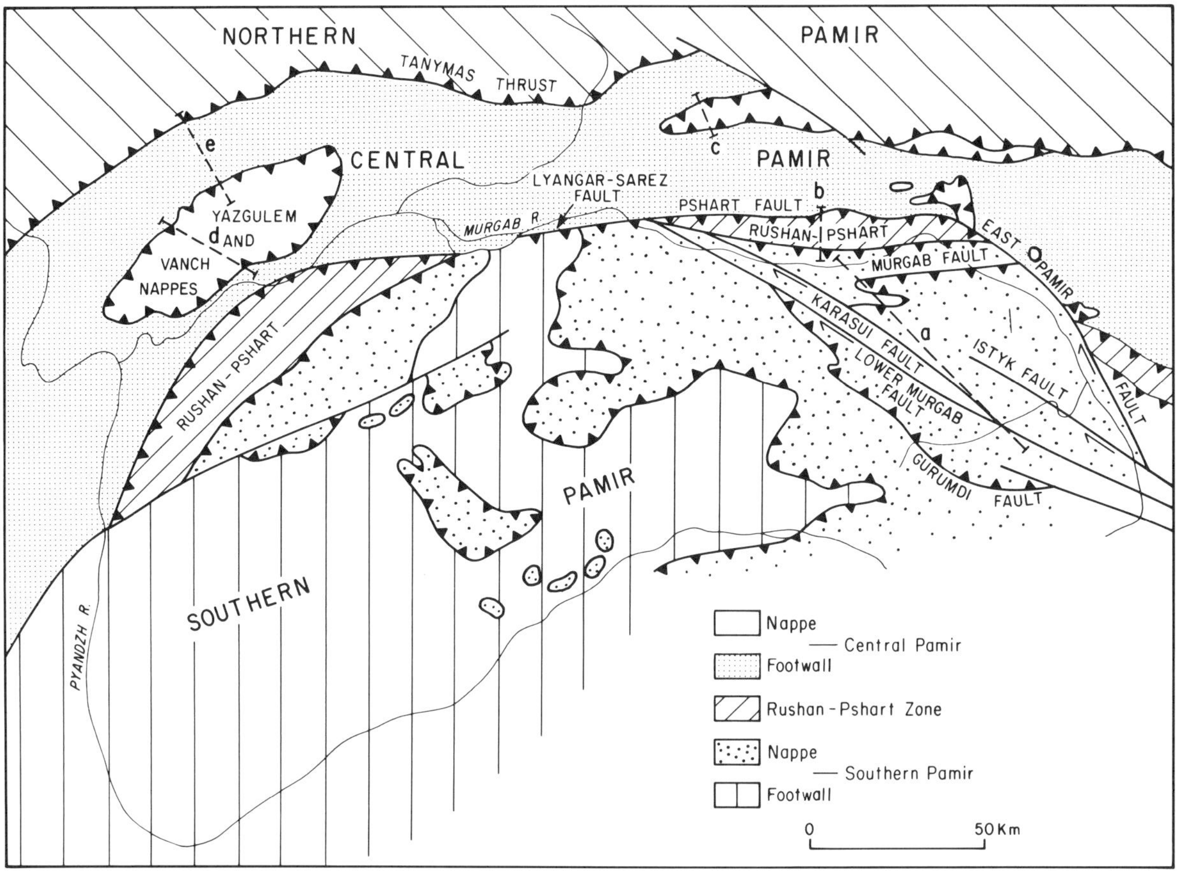

Figure 8. Simplified tectonic map of the Central and Southern Pamir, based on work of Pashkov and Budanov (1990) and Ruzhentsev (1968, 1971) and on the Geological map of the Tadjik SSR and adjoining territories (1984). Note the distribution of nappes and windows in both the Central and Southern Pamir. Lines with teeth indicate major thrust faults; teeth point in dip directions. Lowercase letters show cross sections in Figures 9, 11, 12, and 13.

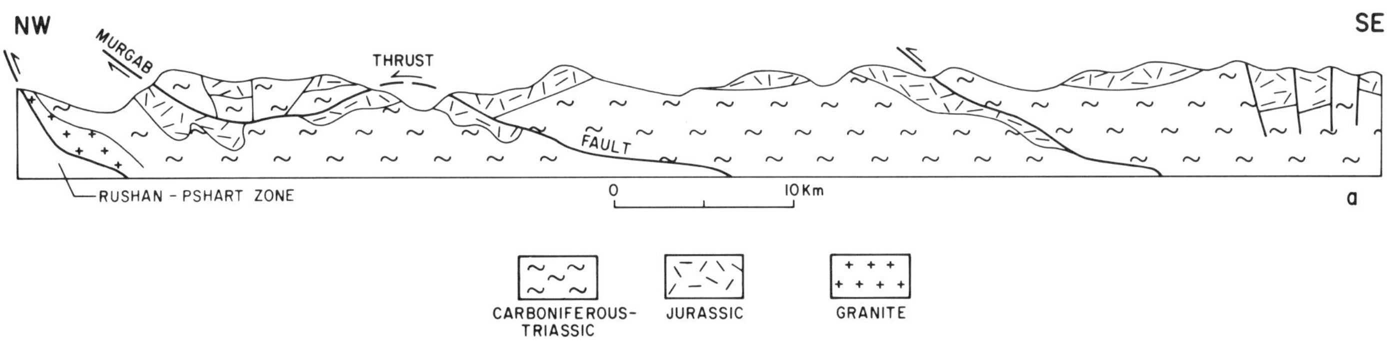

Figure 9. Cross section a across the Murgab thrust fault in the Southeastern pamir. Location of the cross section is denoted by a in Figure 8. Ruzhentsev (1968, pp. 107–114) estimated that 15 to 20 km of displacement occurred on the Murgab fault in Cenozoic time. (Redrawn from Ruzhentsev and Shvol'man [1982].)

Figure 10. Photograph of folded Mesozoic and Paleogene rock of the Southeastern Pamir. View is northwest, parallel to the Aksu-Murgab fault, which diverges from the Pamir-Karakorum fault at the right of the photo (see Figure 3 for location) The dark rocks in the axis of the syncline are Cretaceous/Paleogene red beds. They overlie the well-bedded Jurassic limestone of the Aghil Formation, which dips to the left in the center of the photo. Triassic shale and limestone, also of the Aghil Formation, form the mountains in the background in the center. (Photo by P. Molnar, June 1992.)

4). Peive and others (1964) inferred a cumulative offset of 250 km on this major fault. P. Tapponnier (personal communications, 1992) suggested a north-south component as large as 350 km, corresponding to more than 600 km of strike slip. The presently active, easternmost branch of the Pamir-Karakorum fault passes into a series of grabens and pull-apart basins (Liu and others, 1992) that absorb right-lateral slip. Older right-lateral displacements are clearly revealed by penetrative deformation within the eastern Pamir in China (Brunel and others, 1992b). It appears that the Pamir-Karakorum fault terminates at the east-west–trending Tanymas fault (Fig. 8), a northerly dipping thrust fault separating the Northern and Central Pamir. There are no independent estimates either for the right-lateral slip on the segments of the Pamir-Karakorum fault on the Chinese side or for the Cenozoic thrust slip on the Tanymas fault, but we consider it likely that a significant amount of such slip (tens of kilometers) has occurred. We address this below.

Central Pamir. The footwall beneath thrust sheets emplaced in Cenozoic time consists of late Precambrian (Vendian) to early Paleozoic, late Triassic to Jurassic, and late Cretaceous to Paleogene rock (Pashkov and others, 1981; Ruzhentsev, 1968, 1971, 1990a, 1990b). The thrust sheets exhibit a two-tiered structure. Ordovician to Permian rocks form the lower sheet (called Vanch in the west and Akbaital in the east) and lie on Cretaceous and Paleogene deposits (Figs. 12 and 13). The upper sheet (called Yazgulem in the west and Zortashkol in the east) consists of late Triassic, Jurassic, Cretaceous, and Paleogene rock and lies on both Cretaceous and Paleogene rock deposited on the footwall and on Paleozoic rock of the upper thrust sheet (Figs. 12 and 13).

The internal structure of the footwall is characterized by south-verging recumbent folds that were later deformed by upright folds. The strain revealed by the overturned Vanch

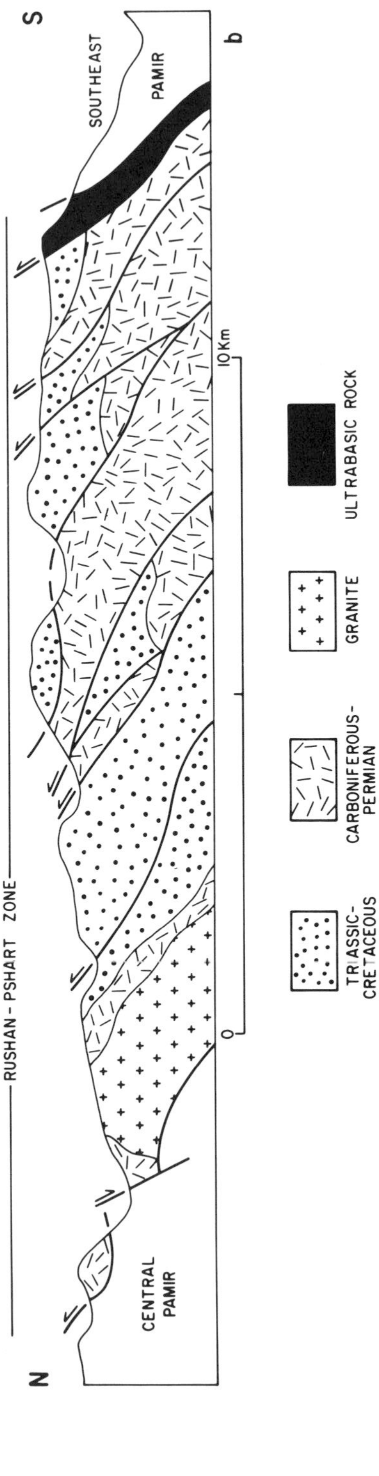

Figure 11. Cross section *b* across the Rushan-Pshart zone. Location of the cross section is denoted by *b* in Figure 8. Imbricate thrust faulting within the Rushan-Pshart zone probably is Mesozoic in age, but the whole Rushan-Pshart zone has been thrust northward onto Tertiary deposits of the Central Pamir. (Redrawn from Ruzhentsev and Shvol'man [1982].)

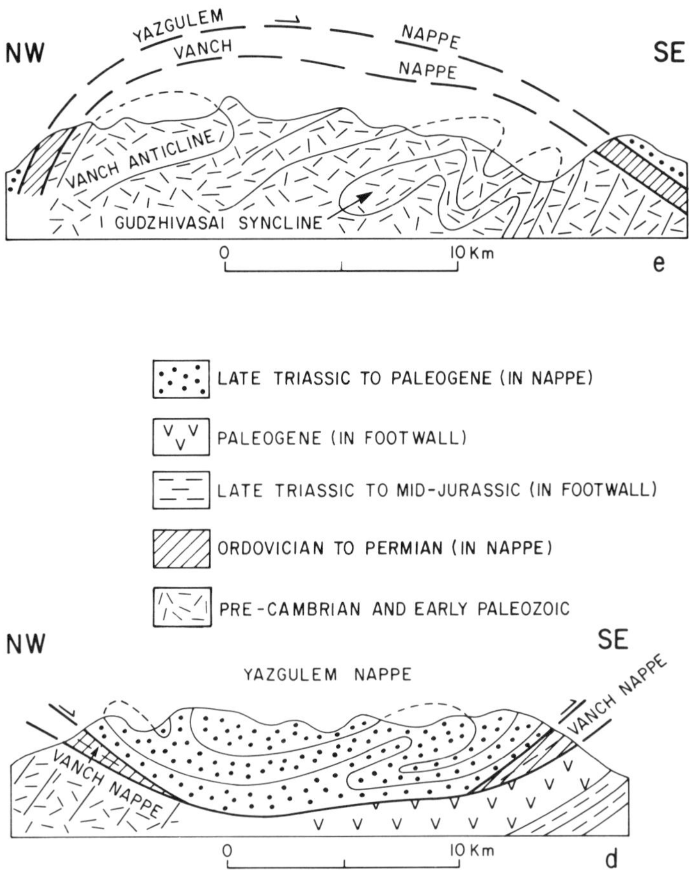

Figure 12. Cross sections across the Yazgulem and Vanch nappes in the Central Pamir, along lines *d* and *e* in Figure 8. The Yazgulem nappe comprises folded Mesozoic and Cenozoic rock deformed into overturned and recumbent folds. To the northwest, both it and the Vanch nappe plunge beneath the Tanymas thrust fault, which dips beneath the Northern Pamir. The relationship of the Yazgulem nappe to the underlying Vanch nappe is not very clear, but both have been thrust over Precambrian to Paleogene rock in the footwall of the Central Pamir. Some of the deformation in the underlying Vanch anticline and Gudzhivasai syncline might be Cenozoic in age. (Redrawn from Ruzhentsev [1990a].)

anticline and by the overturned Gudzhivasai syncline (Fig. 12) indicates shortening of about 10 km; in other cross sections, the shortening associated with the Vanch anticline reaches 20 km (Ruzhentsev, 1971, p. 18). These folds formed in rocks of late Precambrian (Vendian) and early Paleozoic age, and the times of deformation are not clear. Overturned folds are also well developed in Triassic and Jurassic sedimentary rock in the footwall, but elsewhere these deposits form a continuous stratigraphic sequence with early Cretaceous rock, which places a lower limit on the age of folding. Ruzhentsev (1971, p. 32) suggested that the folding occurred before the deposition of late Cretaceous sedimentary rock that overlies the early Cretaceous rock on an angular unconformity. Possibly some of the folding of the footwall occurred later.

The Vanch, Akbaital, and Zortashkol nappes actually consist of several tectonic slices (Fig. 13). The Mesozoic and Cenozoic rock of the Yazgulem nappe form large, south-verging overturned and recumbent folds (Fig. 12). The root zone of the nappe lies at the northern edge of the Central Pamir and is concealed under the Tanymas thrust fault dipping beneath the Northern Pamir. Thus, as noted above, the Central Pamir has been underthrust beneath the Northern Pamir. Ruzhentsev (1990b) reported a cumulative displacement of the thrust sheets within the Central Pamir of about 80 km. This value, however, ignores not only shortening of the thrust sheets by folding and later disruption by strike-slip faulting but also underthrusting on the Tanymas fault, which as noted above might exceed tens of kilometers.

If we take 100 km as a lower bound on the Cenozoic convergence between the southern edge of the Central Pamir and the Northern Pamir and 240 km across the Southern Pamir and Rushan-Pshart zone, we obtain 340 km, or more, of crustal shortening of the Pamir in Cenozoic time. This sum assigns shortening to the Cenozoic era in some places where earlier shortening cannot be eliminated. It includes estimates made at different localities. Thus, it ignores the possibility that internal bending and rotation of blocks about vertical axes make local estimates unrepresentative of the average shortening across the entire region. But it also ignores deformation in regions where Cenozoic rock is absent. Thus assigning a meaningful uncertainty to this number is difficult. We proceed with the assumption that it gives an adequate approximation to the amount of internal shortening: more than 300 km.

Convergence of the outer margin of the Pamir with the South Tien Shan

Not only are the tectonic belts within the Pamir deflected with respect to their continuations to the east and west and deformed internally, but the entire Pamir has been displaced northward with respect to the South Tien Shan. This northward displacement is reflected by both displacements and deformation of rock deposited in the Tadjik Depression and the Pamir-Alai region. Cretaceous and Paleogene sedimentary rock crops out widely in "Tadjik sedimentary basin," which includes the Gissar (Hissar) and Alai ranges along the South Tien Shan, the Tadjik Depression, and the Pamir-Alai region. The variations in facies and thicknesses indicate that this sediment was deposited in an east-west–trending basin that was bounded by highlands on both the north and the south. Suvorov (1968) showed that the east-west–trending facies zones in the Tadjik Depression and the Pamir-Alai range have been displaced with respect to one another. We have augmented his data with newer observations. Here we use them to place a lower bound on the distance that the Pamir has penetrated into the rest of Eurasia. In addition, intense shortening of the sedimentary rock of the Tadjik sedimentary basin attests to additional convergence of the Pamir with respect to the South Tien Shan. We

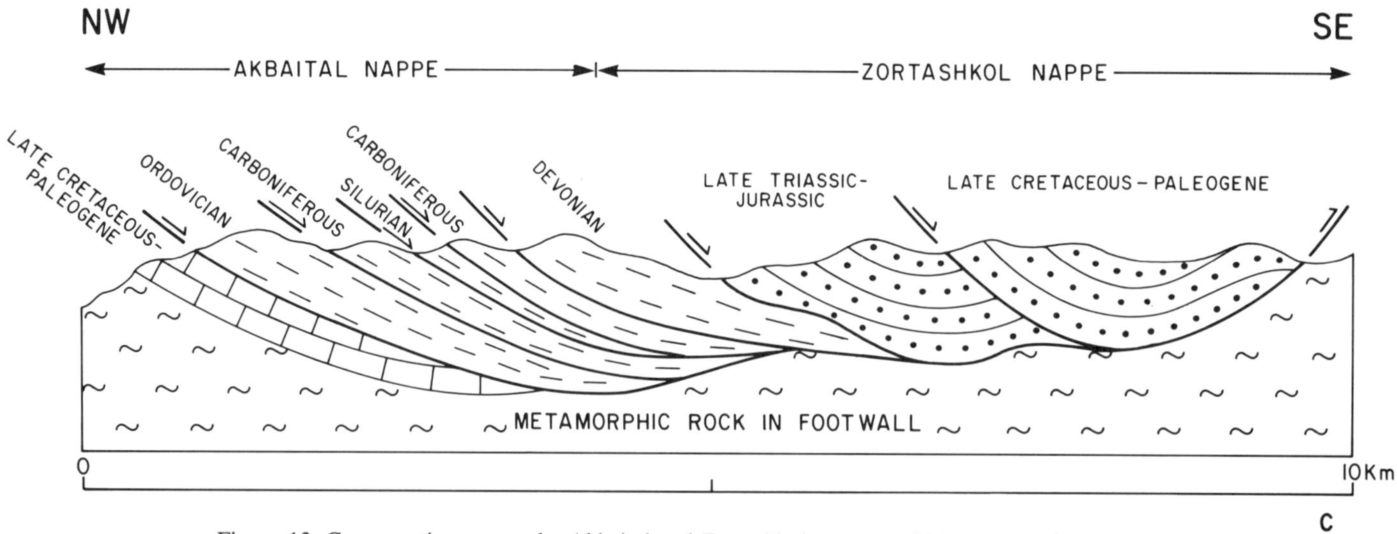

Figure 13. Cross section across the Akbaital and Zortashkol nappes, which consist of several tectonic slices, in the Central Pamir, along *c* in Figure 8. These nappes seem to root in the Tanymas fault between the Northern and Southern Pamir. (Redrawn from Ruzhentsev [1990b].)

address the displacements of facies belts and the shortening of sedimentary rock separately.

Displacements of Cretaceous and Paleogene facies belts from the Tadjik sedimentary basin. It is convenient to separate the Tadjik sedimentary basin into two domains separated by the Vakhsh and Trans-Alai overthrusts (Figs. 14, 15, and 16): a "Tien Shan" domain, which includes the South Tien Shan and the central and western parts of the Tadjik Depression, and an "Outer Pamir" domain, which contains the Pamir-Alai region and the eastern part of the Tadjik Depression. As noted above, paleomagnetic results from early Cretaceous and Paleogene rocks of the outer Pamir domain (from the Trans-Alai, Peter the First, and Darvaz ranges) show large declination anomalies. Thus, the strikes of facies zones in these areas have been bent and distorted since Paleogene time. In contrast, paleomagnetic results from early Cretaceous rocks from the Alai Range of the South Tien Shan and sparse results from Paleogene samples from the central part of the Tadjik Depression reveal no resolvable declination anomalies (Bazhenov and Burtman, 1990). Thus, the strikes of facies zones of the Tien Shan domain have not been much distorted.

Early Cretaceous. Differences both in grain sizes and in the extent of marine incursions allow mapping of facies zones in the largely nonmarine, early Cretaceous sequence. Apparently, early Cretaceous sediment was not deposited in parts of the South Tien Shan and in what is now the northern part of the Pamir (Zones I and VII in Fig. 14). Along the southern margin of the South Tien Shan, sandstone and conglomerate containing clasts of Paleozoic rock from the South Tien Shan attest to early Cretaceous denudation. This conglomerate grades southward into sandstone (zone II in Fig. 14). The southern and eastern parts of this facies zone contain clay and limestone with late Albian marine fossils near the top of the sequence. The thickness of early Cretaceous deposits increases from north to south across this zone from 20 to 400 m. South of facies zone II, finer-grained material is represented by nonmarine sandstone and clay (zone III in Fig. 14). The upper part contains marine clay with early and late Albian fossils. The thickness of the entire early Cretaceous section ranges from 200 to 500 m. The sequence farther south (in zone IV in Fig. 14) is similar but thicker (500 to 800 m), and marine conditions prevailed for a longer period. The marl and limestone contain Hauterivian to Albian fossils.

These zones, in which the thicknesses and the proportion of marine deposits increase southward, strike roughly east-west across the western and central parts of the Tadjik Depression, until they are abruptly truncated by the Vakhsh and Trans-Alai thrust belt (Fig. 14).

Some 800 to 1,300 m of nonmarine sandstone, with rare cross-bedding and clay but abundant plant remnants, define the northwestern facies zone V (Fig. 14) of the outer Pamir domain and apparently were deposited in the southern part of the Tadjik sedimentary basin. In contrast, 400 to 1,400 m of coarse-grained sandstone and conglomerate characterize the early Cretaceous sequence in the eastern and southeastern parts of the outer Pamir domain (VI in Fig. 14). The relatively coarse material and the composition of the clasts suggest the zone VI was closer to the southern edge of the basin than was zone (Fig. 14). Moreover, both grain sizes and compositions of the Pamir domain differ markedly from those in Tien Shan domain. Thus, their current close juxtaposition implies that the facies zones in the outer Pamir domain have been displaced northward past those of the Tien Shan domain and rotated with respect to them.

Sediment of early Cretaceous age can also be found in the Central Pamir (VIII in Fig. 14). Both coarse-grained conti-

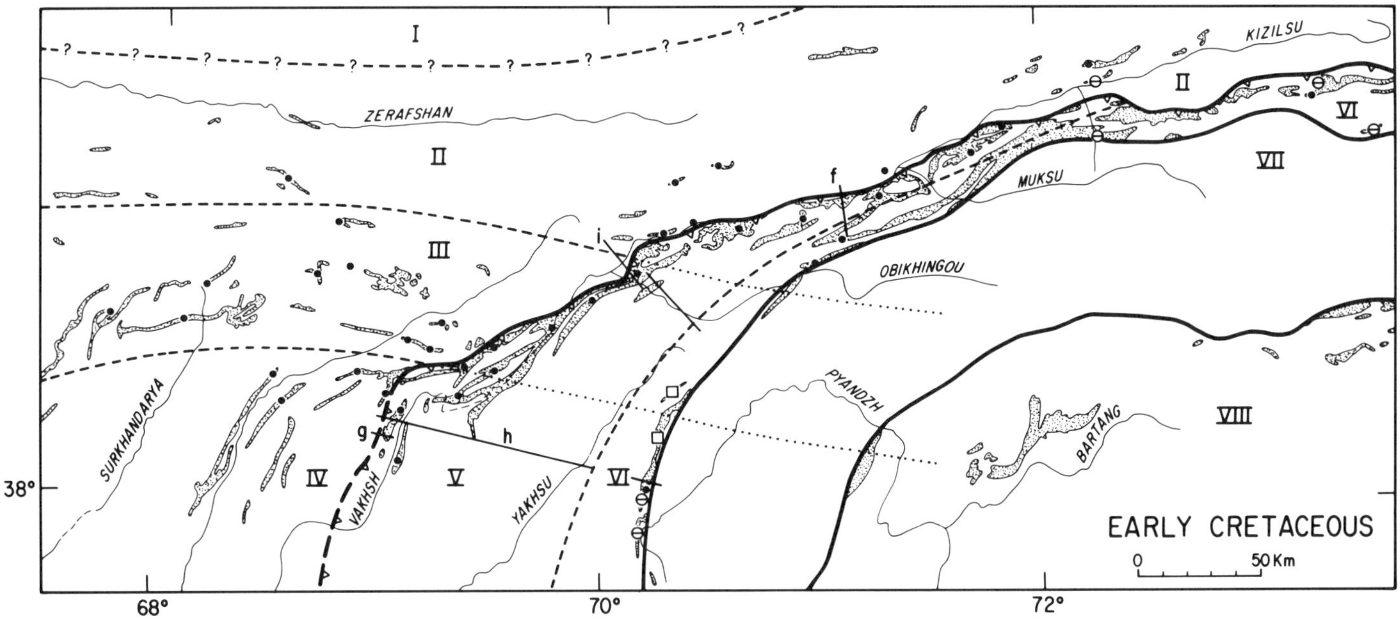

Figure 14. Map of early Cretaceous facies zones in the Tadjik Depression and Pamir-Alai region (after Suvorov [1968], with additions). Throughout the region shown, early Cretaceous sedimentary rock, whose outcrops are shown by shading, is dominated by nonmarine deposits, but variations in grain sizes of this material and in the distribution and ages of marine deposits near the top of the sequence allow several facies zones to be delineated. Dashed lines separate regions where different facies zones can be defined, and continuations of some of these dashed lines are projected as dotted lines beneath the overthrusted Pamir mass. The thick line with triangles shows the Vakhsh and Trans-Alai overthrusts. The thick lines farther south show the Darvaz-Karakul and Tanymas faults. Zones I and VII encompass areas where erosion occurred in early Cretaceous time in the south Tien Shan and in the Northern Pamir, respectively. Zone II shows the distribution of nonmarine, coarse-grained deposits, containing fragments of Paleozoic rocks of the South Tien Shan. Zone III shows the distribution of largely nonmarine, fine-grained deposits but with layers of marine deposits of Albian age. Zone IV shows the distribution of largely nonmarine, fine-grained deposits but with layers of marine sediment of Hauterivian, Aptian, or Albian age. Zone V contains only nonmarine, fine-grained deposits, and Zone VI contains both fine- and coarse-grained nonmarine deposits with fragments of the rocks of the Northern Pamir. Early Cretaceous sediment is rare in Zone VII. Zone VIII contains coarse-grained continental and lagoonal debris. Closed circles indicate data from Suvorov (1968, Fig. 71), open squares are from Kariev (1977), and open circles with lines through them are Burtman's observations. Also shown are the locations of cross sections f, g, h, and i, in Figures 17, 19, and 20.

nental and lagoonal marine deposits were deposited in a marine area separated from that of the Tadjik Depression by an emergent area now lying in the Northern Pamir (Shvol'man, 1977, pp. 31–72).

Late Cretaceous. The relatively homogeneous sediment of late Cretaceous age, deposited in a marine or lagoonal environment, makes separating facies zones more difficult than for older or younger material. Nevertheless, the east-west trend of the facies zones and their truncation at the Vakhsh Overthrust are clear.

The northern part of the Tadjik Depression and the South Tien Shan (zone II in Fig. 15) contains 200 to 700 m of late Cretaceous sandstone, clay, gypsum, and limestone. Conglomerate, present near the northern edge of this zone, attests to a nearby emergent region undergoing denudation. Correspondingly, limestone, absent from the northern part, constitutes most of the sequence in the southern part of zone II.

In the more southern belt of the Tien Shan domain (III in Fig. 15), clay dominates a 700- to 1,000-m-thick sequence that also contains beds of limestone, sandstone, and gypsum. The transition between the two zones trends east-southeast to east-west.

Limestone and marl dominate the Campanian-Maastrichtian sequence in the outer zone of the Pamir, but in the lower part of the section limestone alternates with clay, sand, and gypsum. Although the thickness varies from 500 to 1,300 m, systematic differences in thicknesses allow two zones in this domain to be distinguished. Thicknesses exceed 1,000 m in the northwestern zone (IV) but are less than 1,000 m in the southeastern zone (V in Fig. 15).

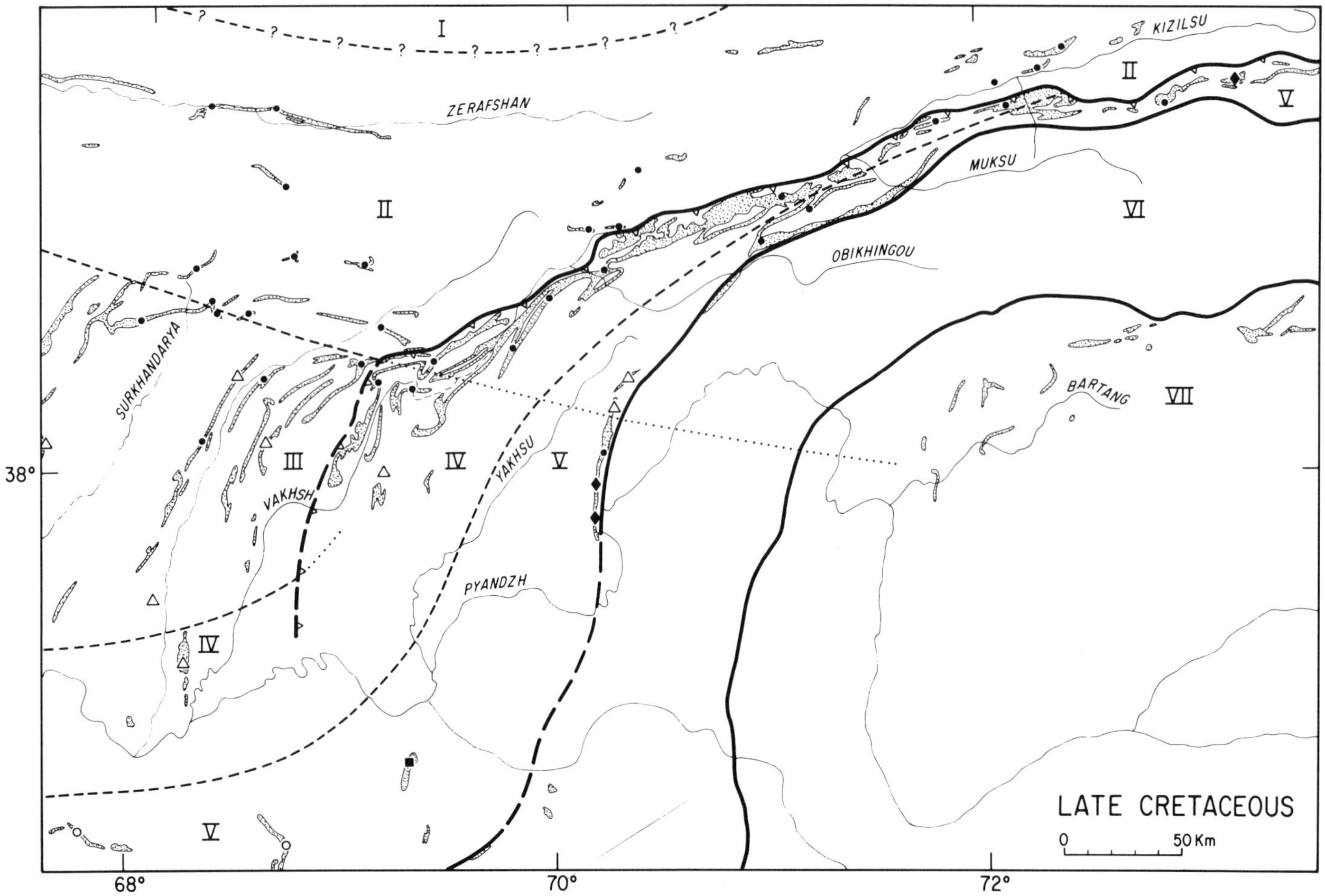

Figure 15. Map of late Cretaceous facies zones in the Tadjik Depression and Pamir-Alai region (after Suvorov [1968], with some changes and additions). As in Figure 14, regions with outcrops are shaded. Zone I shows the area where erosion occurred in the South Tien Shan during late Cretaceous time. Zone II contains limestone, gypsum, clay, and sandstone with beds of conglomerate. Zone III contains clay with beds of limestone, gypsum, and sandstone. Zone IV contains limestone and marl, with beds of gypsum and clay, and with a total thickness of 1,000 to 1,300 m. Zone V is similar to zone IV, but the thickness is 500 to 1,000 m. No late Cretaceous sedimentary rock has been found in zone VI. Marine and lagoonal deposits crop out in Zone VII. The remainder of the legend is as in Figure 14. Closed circles indicate data from Suvorov (1968, Fig. 72), closed square from Bratash and others (1970. pp. 115–136), open circles from Dronov (1980, pp. 211–217), open triangles from Dzhalilov (1971, pp. 30–149), and closed diamonds from Burtman's observations.

It appears that the sequence of sediment deposited in the interior of the Tadjik sedimentary basin also crops out in the outer Pamir domain. Correspondingly, the late Cretaceous position of the southern border of the basin cannot be reconstructed with certainty, because there are no late Cretaceous deposits within the Northern Pamir (VI in Fig. 15). Nevertheless, the composition of Cenomanian sedimentary rock at the eastern end of the Trans-Alai range suggests that the southern border of the basin is not far south of this locality. Sandstone and mudstone, suggesting a proximal terrigenous source, make up much of that sequence. Moreover, some beds of sandstone show cross-stratification and conglomerate interbeds with small pebbles (Leonov, 1961, p. 36). Conglomerate is also present in the late Cretaceous section in the region between the Pyandzh and Yakhsu rivers near the Darvaz fault (V in Fig. 15) (Dzhalilov, 1971, pp. 134–149).

Late Cretaceous carbonate sediment, of marine and lagoonal facies, was also deposited within the Central Pamir in Campanian and Maastrichtian time (VII in Fig. 15). The area surrounding the Bartang Valley, however, apparently underwent denudation in Cenomanian time. Conglomerate with a basal unconformity overlies early Mesozoic rock and grades upward

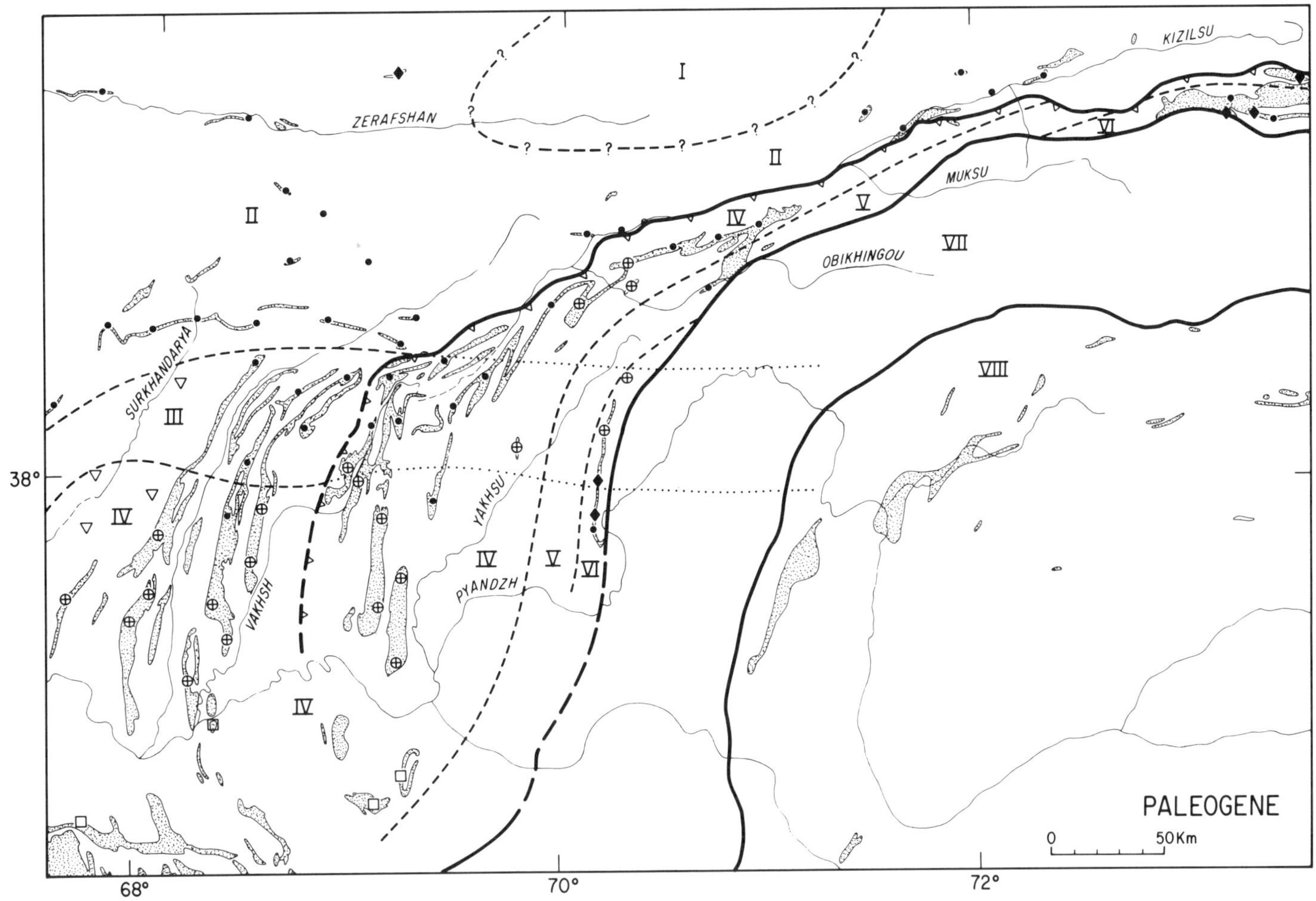

Figure 16. Map of Paleogene facies zones in the Tadjik Depression and Pamir-Alai region (after Suvorov [1968] with some changes and additions). Zones I and VII show areas where erosion apparently occurred in Paleogene time. Zones II and VI contain marine deposits in the lower part of the sequence and nonmarine deposits in the upper part. The thicknesses of marine deposits range from 40 to 350 m. Zones III and V are similar to zones II and VI, but thicknesses of marine deposts range from 350 to 500 m. In zone IV, the entire Paleogene sequence is marine, and its thickness ranges reach 1,300 m. Coarse-grained terrigenous material was deposited in shallow water in zone VIII. The remainder of the legend is as in Figure 14. Closed circles indicate data from Suvorov (1968, Fig. 72), open squares are from Bratash and others (1970, pp. 115–136), open circles with pluses in the center are from Davidson and others (1982, pp. 41–76) open downward-pointing triangles are drill holes reported by Minakova and others (1975, pp. 12–47), and closed diamonds indicate Burtman's observations.

into Campanian and Maastrichtian carbonate (Shvol'man, 1977, pp. 31–72). Thus, the Tadjik sedimentary basin seems to have extended southward over the North and Central Pamir and may have included a much larger area than it did earlier and later.

Paleogene. The Paleogene facies zones reveal an approximate axial symmetry of the Tadjik sedimentary basin with marine conditions prevailing for the longest time in the center of the basin. Along the margins of the basin (in zones II and VI in Fig. 16), marine sedimentation occurred in the Paleocene and Eocene epochs; 40 to 350 m of marine and lagoonal deposits are represented by limestone and marl, with beds of gypsum, clay, and sandstone. Then, nonmarine sandstone and clay were deposited in Oligocene time. In the more inner parts of the basin (zones III and V in Fig. 16), marine sedimentation continued until early Oligocene time, with thicknesses of marine deposits ranging from 350 to 550 m. In the central part of the basin (belt IV in Fig. 16), the entire Paleogene sequence, up to 1,300 m thick, consists of marine deposits. Note that zones II, III, and VI are abruptly truncated at the Vakhsh Overthrust (Fig. 16).

Paleogene sediment is also found in the internal parts of

the Pamir (VIII in Fig. 16). This coarse-grained continental and volcaniclastic material was deposited in shallow water in a basin within the Central Pamir (Shvol'man, 1977, pp. 31–72).

Summary. The truncation of the east-west–trending Cretaceous and Paleogene facies zones in the western and central Tadjik Depression by those in the outer arc of the Pamir corroborates the inference of large-scale overthrusting of the Pamir arc onto the Tien Shan domain. These east-west–striking facies zones terminate abruptly at the Vakhsh and Trans-Alai overthrusts. Thus, it seems reasonable to assume that they continue eastward in the subsurface beneath the thrust sheets in the eastern part of the Tadjik Depression. Because the southwesternmost facies zones in the unrotated, western and central parts of the Tadjik Depression are not represented in the outer Pamir domain, a definitive estimate of the northward displacement of the Pamir with respect to the South Tien Shan cannot be given, but that amount must be at least 200 km.

Cenozoic crustal shortening within the outer zone of the Pamir. The configuration of the Pamir arc implies that the original extent of the Tadjik sedimentary basin was much larger than the present area occupied by the Tadjik Depression and Pamir-Alai region. Only Mesozoic and Cenozoic sedimentary rock crops out in the area west of the Pamir-Alai region in the Tadjik Depression and east in the Yarkand Depression at the western end of the Tarim Basin (e.g., Gubin, 1960; Leith, 1982; Leonov, 1961; Sinitsyn, 1957; Zakharov, 1958). A wide sedimentary basin, the Tadjik-Yarkand basin, seems to have embraced the entire region now occupied by the Pamir up to the late Eocene epoch. Late Cretaceous and Paleogene marine sediments were deposited within this basin. Then, in late Eocene time, soon after the beginning of the collision between India Eurasia, not only did the sea retreat but isolation of the Tadjik and Yarkand parts of the basin began. A 4- to 6-km-thick sequence of coarse Neogene sediment was deposited on the Cretaceous and Paleogene sequence of the Pamir-Alai region and in the Tadjik and Yarkand depressions, and accumulation of this material occurred with the growth of the Pamir and South Tien Shan.

In the Pamir-Alai region, the facies zones of the southern (Pamir) side of the Tadjik-Yarkand basin are juxtaposed adjacent to those of the northern (South Tien Shan) side (Figs. 14, 15, and 16), and a large part of this region is occupied by sedimentary rock that was deposited on the southern side of the basin. These deposits mark the outer zone of the Pamir, which has been thrust onto Mesozoic and Cenozoic rock of the southern edge of the South Tien Shan along the Vakhsh and Trans-Alai overthrusts (Figs. 3 and 17), which appear to define the outer boundary of the Pamir.

The different layers of rock in the outer zone of the Pamir are characterized by very different rheological properties (e.g., Hamburger and others, 1992; Leith and Alvarez, 1985; Leith and others, 1981; Skobelev and others, 1988; Zakharov, 1964). The Triassic and most of the Jurassic sedimentary rocks are separated from the overlying Cretaceous sequence by horizons of Tithonian gypsum and salt. This weak layer has a thickness of several hundred meters and serves as the boundary between an essentially intact basement and a cover consisting of Cretaceous and Cenozoic deposits, which has been compressed into disharmonic folds and detached from the underlying basement. The Tithonian gypsum and salt have been carried with the cover along thrust (and normal) fault surfaces in the Trans-Alai, but in Tadjik Depression they form salt domes.

The Cretaceous and Cenozoic cover is itself inhomogeneous. In much of the region, it contains (1) a mildly ductile layer of early Cretaceous sandstone 400 to 1,400 m thick; (2) a more ductile layer of late Cretaceous and Paleogene limestone, marl, gypsum, and clay 500 to 2,000 m thick; and (3) another mildly ductile layer of Neogene conglomerate and sandstone, more than 5,000 m thick. As a result, folds within the cover are also disharmonic (Fig. 18) (Gubin, 1960, pp. 197–212; Skobelev and others, 1988).

The folded structure seen in the outer zone of the Pamir seems to have formed near the end of the Cenozoic period. A pronounced angular unconformity separates Pleistocene sediment from the underlying, intensely deformed Cretaceous, Paleocene, and Neogene sedimentary rock, but Quaternary folding is clear in some areas. Regional uplift of the Pamir also is said to have occurred at this time (Belousov, 1976; Yermilin and Chigarev, 1981), with the formation of the regional topography and slip on faults. One of us (P. M.), however, thinks that much of the evidence for Quaternary relief is probably a result of erosion resulting from global climate change and may not be due to a change in tectonics in late Neogene time (Molnar and England, 1990).

Because the folds in the Pamir-Alai region are disharmonic and complicated, it is impossible to determine reliably the amount of crustal shortening from the internal deformation of these rocks seen at the surface, in a zone only 15 to 25 km wide. Direct measurements of displacements on individual faults are relatively small. In the eastern part of the Trans-Alai Range, Cretaceous and Paleogene rock have been thrust onto Neogene deposits with a visible displacement reaching 10 km (Gubin, 1960, pp. 243–244). In some exposures it is possible to see that this fault surface dips south at an angle of 10 to 20°, but nearby on a neighboring thrust fault Paleogene rocks have been thrust over early Quaternary deposits on a steeper fault (Nikonov and others, 1983, p. 81). Farther west in the central part of the Trans-Alai Range, Cretaceous rock on the north slope has been thrust on a flat surface over Neogene deposits. Nikonov and others (1983, p. 79) described olistostromes with olistoliths of rock from the thrust sheet in Pliocene sediment. Several klippen of this nappe imply a minimum displacement of 8 km. To the north of the main thrust fault, there are Quaternary thrust and reverse faults. Along one of them, Neogene rock has been thrust onto late Pleistocene deposits (Nikonov and others, 1983, pp. 56, 79–82). In the western part of the Trans-Alai Range,

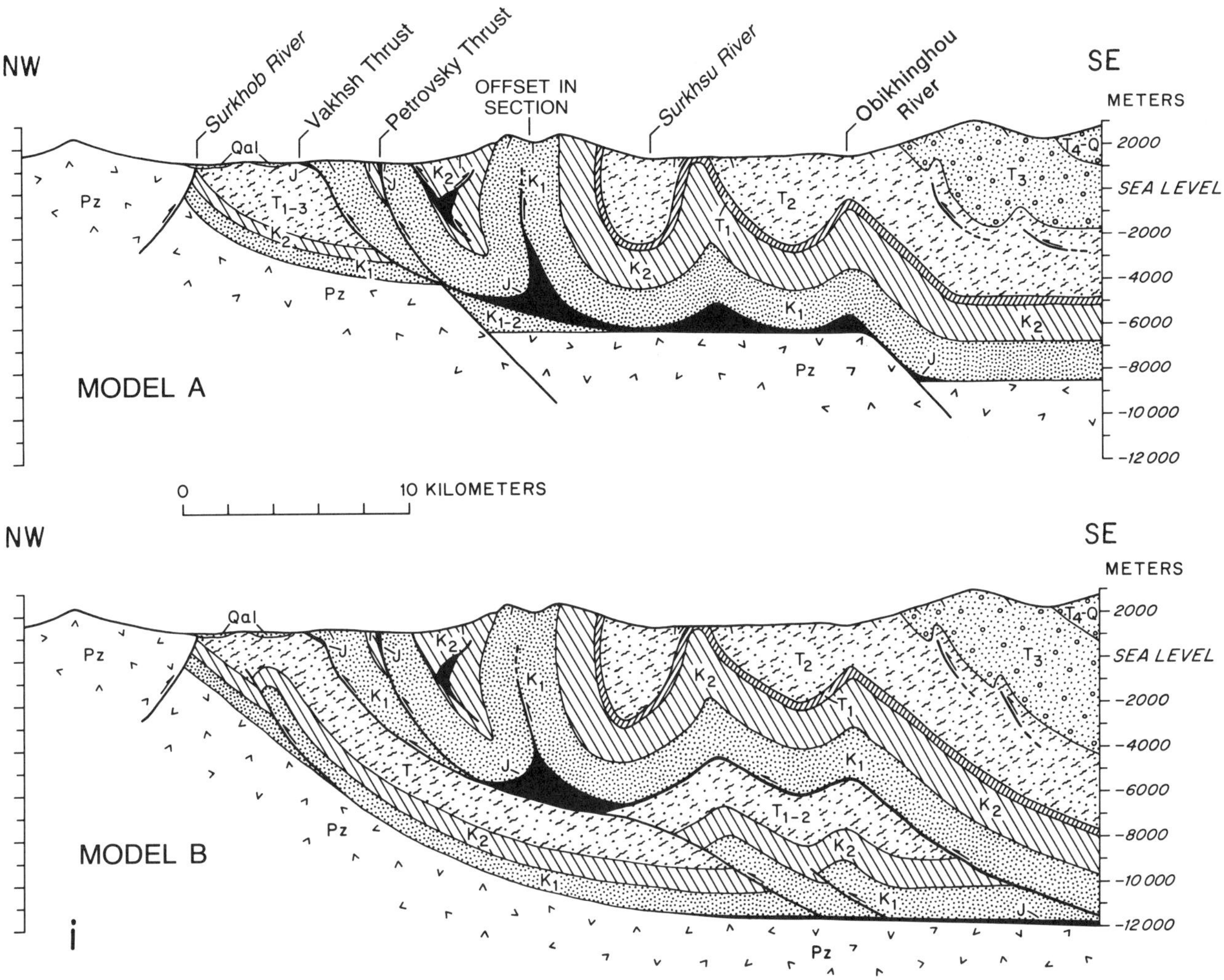

Figure 17. Plausible balanced cross sections, based on the same geologic observations, across the Peter the First Range near Garm (after Hamburger and others 1992). Location of cross section denoted by *i* in Figure 14. The following abbreviations indicate the corresponding units: Pz—Paleozoic crystalline rock, J—Jurassic evaporite, K_1—Early Creataceous largely sandstone and mudstone, K_2—Late Cretaceous largely limestone, T_1—Tertiary (Paleogene) limestone and evaporite, T_2—Tertiary (Miocene) mudstone and sandstone, T_3—Tertiary (Mio-Pliocene) and T_4-Q—Tertiary (Plio-Quaternary) sandstone and conglomerate, Qal—Quaternary alluvium. Note the greater thickness of Mesozoic sedimentary rock in the thrust sheets than in the section deposited on the Paleozoic rock near the Surkhob River. The top cross section begins with the assumption (attributed to Leith [1985]) that the southward increase in thicknesses of Mesozoic strata is abrupt and is associated with crustal thinning and normal faulting at the edge of a marginal basin. The lower cross section, favored by Hamburger and others (1992), is based on the assumption that the differences in thicknesses result from more gradual thickening and large horizontal transport of the thrust sheets. In both cross sections, Cretaceous and younger deposits have been detached along Jurassic salt units (black).

Cretaceous rock from the outer zone of the Pamir has been thrust completely over the Paleogene and Neogene rock deposited on the Paleozoic rock at the southern margin of the South Tien Shan, implying at least 15 to 25 km of shortening.

Farther to the southwest, the Vakhsh Overthrust follows the northern foot of the Peter the First Range along the Surkhob River. On the northern bank of this river, Paleozoic rock of the South Tien Shan crop out, and remnants of the overlying Cretaceous and Paleogene section are preserved; on the northern slope of the Peter the First Range, there is a system of thrust sheets containing Cretaceous rock (Fig. 17). The surface of the thrust fault is marked by exposures of late Jurassic gyp-

Figure 18. Photographs of disharmonically folded Cretaceous rock in the Peter the First Range. A, Close-up of "Kamenii Tsvetok" ("the stone flower"), a box fold in Paleogene limestone and mudstone. View is southwest along the westerly flowing reach of the Obikhingou in the middle of cross section *i* (Fig. 17). See also Hamburger and others (1992) for another view of this fold. B, A second example of folded Paleogene and Cretaceous rock in the Peter the First Range. View is northeast along the axis of the range, in the opposite direction from that in A. (Photos by P. Molnar, July 1974 [A] and November 1973 [B]).

sum (Fig. 19). At the foot of the range, Cretaceous rock has been thrust over Pleistocene deposits (Hamburger and others, 1992; Lukk and Shevchenko, 1986; Nikonov and others, 1983, pp. 82–83; Skobelev and others, 1988). As discussed below, this fault has been active in Holocene time (Trifonov, 1983, pp. 82–87).

This same imbricate structure characterizes the Vakhsh Overthrust to the west, along the foot of the Vakhsh and Surkhu ranges. The lower strand of the thrust fault dips gently to the south with a minimum displacement of 2 to 3 km. Jurassic and Cretaceous sedimentary rock has been thrust onto Neogene and older rock of the South Tien Shan (Gubin, 1960, pp. 240–242; Leith and Alvarez, 1985). Farther west, the Vakhsh Overthrust bends to the south. Although concealed by Quaternary sediment in some areas, the structure is known from drill holes (Figs. 19 and 20).

Thus, the structure along the Vakhsh and Trans-Alai overthrusts attests to significant, localized displacement of the outer zone of the Pamir over the southern edge of the South Tien Shan and the Tadjik Depression, but offsets of rock units defined chronostratigraphically yield only minimum thrust displacements that are not very useful as constraints in evaluating the magnitude of overthrusting.

The variations in dimensions of the outer zone of the Pamir, however, suggest a substantial amount of internal shortening of this zone. First, the width of the outer zone of the Pamir, about 100 km in the cross section in Figure 20, decreases to only 15 to 25 km in the Peter the First Range (Fig. 17). Ignoring the roughly 35 km of overthrusting on the Vakhsh overthrust in the profile in Figure 20, the total shortening of the outer zone of the Pamir measured from offsets and folding of the Paleogene-Cretaceous interface is at least 35 km. This implies that the width of the outer zone along the profile in Figure 20 was once 135 km. Recognizing that the outer zone of the Pamir has been rotated, let us assume that the same width characterized the northern part of the outer zone, which is now only 15 to 25 km wide. The difference in widths, between 135 km and 15 to 25 km, suggests that the northern margin has been shortened by about 115 km, not the mere 15 to 25 km that can be determined from offsets in the Peter the First Range. This shortening must be added to the 200 km or more of northward translation inferred from the offset facies boundaries. Thus, ignoring details, we infer that the northern margin of the Pamir has advanced over the Tadjik sedimentary basin 300 km, or more, since Paleogene time. Note also that additional Cenozoic northwest-southeast shortening has occurred within the Tadjik Depression to the northwest of the section in Figure 20. Hence, the Tadjik sedimentary basin must have extended eastward 100 km or more into the territory now occupied by the southwestern Pamir.

It is instructive to consider the fraction the convergence between India and Eurasia since their collision that has been absorbed in the Pamir. Cenozoic convergence across the Pamir-Punjab syntaxis includes both 340 km (call it more than 300 km) of internal crustal shortening within the Pamir and 315 km of overthrusting of the Pamir onto the area farther north. Thus, the sum exceeds at least 600 km (and probably 650 km). We are aware of only very limited study of the eastern Hindu Kush of northeastern Afghanistan and of the western Karakorum in northern Pakistan and western Tibet, which

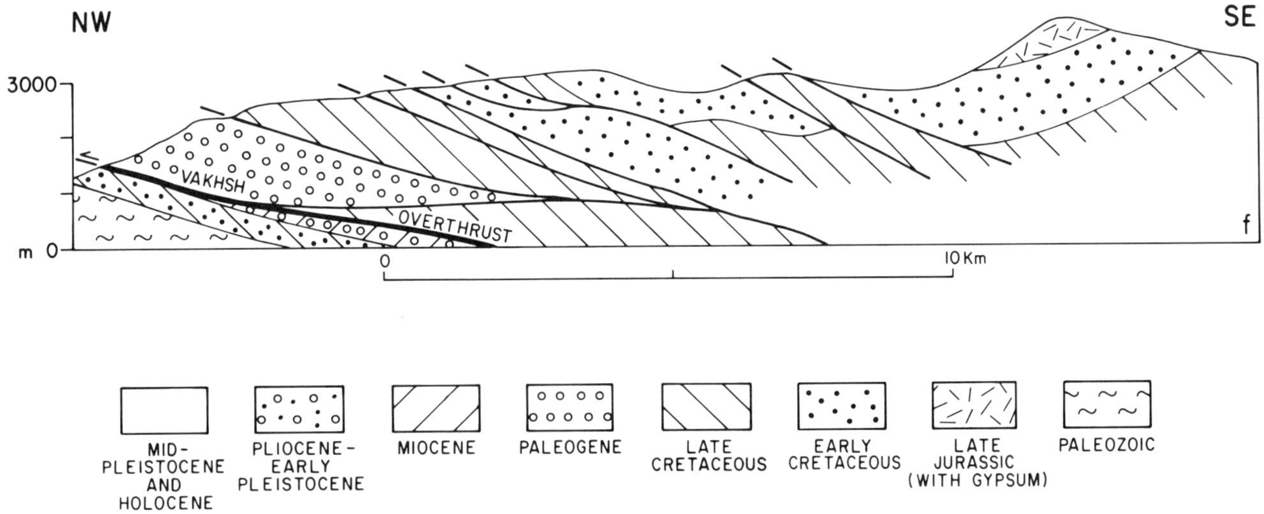

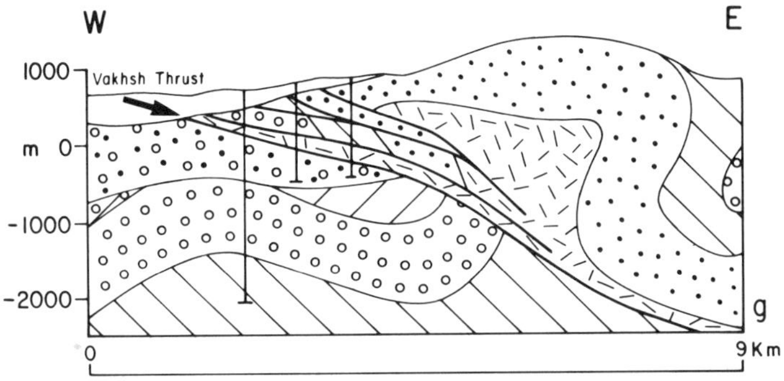

Figure 19. Cross sections *f* and *g* across the Vakhsh and other overthrust faults. Locations of cross sections are denoted by *f* and *g* in Figure 14. Cross section *f*, compiled from data of Skobelev (1977) and Skobelev and others (1988), shows complicated imbricate thrust faulting of Mesozoic and Cenozoic sedimentary rock of the Peter the First Range north-northwestward onto the South Tien Shan. This cross section trends 170°. East-west cross section *g*, from Varentsov and others (1977, p. 64) crosses the Vakhsh Overthrust in the Tadjik Depression. Its location and orientation are constrained by drill holes (vertical lines). Notice that large magnitudes of thrust faulting are possible but cannot be confirmed by these cross sections.

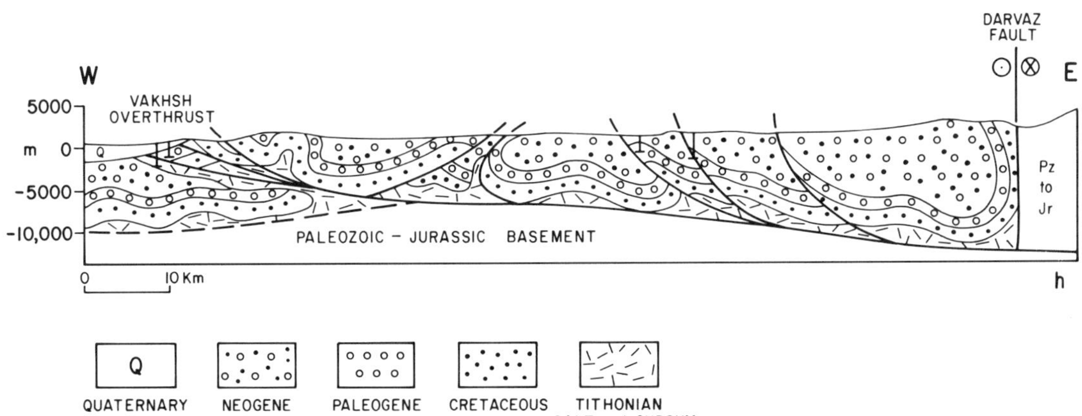

Figure 20. Geologic cross section *h* across the outer zone of the Pamir. Location of cross section is denoted by *h* in Figure 14. This was constructed from geological and geophysical data presented by Bekker and others (1974a, 1983). Note the large amount of shortening by thrust faulting in the sedimentary cover of the outer zone of the Pamir and the large displacement on the Vakhsk Overthrust. The rock east of the left-lateral Darvaz strike-slip is mostly pre–middle Jurassic "basement" and moves away, northward, with respect to that on the west side.

lie south of the Pamir. Gaetani and others (1990) reported evidence of Mesozoic folding and thrusting faulting in northernmost Pakistan, probably of early Cretaceous age, but no subsequent deformation. Consequently, we cannot estimate the Cenozoic shortening of this region. Farther south, in Kohistan, Pakistan, a hypothetical restoration of the terrain between the Indus suture zone and to south of the Main Boundary fault of the Himalaya by Coward and others (1987) yields an estimate of 470 km of north-south shortening. The sum of these numbers suggests a total convergence approaching, if not exceeding, 1,100 km. Since the collision between India and Eurasia, these two continental masses have moved toward one another 1,800 ± 400 km since 40 Ma and roughly 3,000 km since 50 Ma (Molnar and Tapponnier, 1975). The minimum of 1,100 km of shortening falls short of these amounts and implies that some convergence has not been taken into account. Some material may have been translated (extruded) laterally eastward or westward out of India's northward path, and some may have been subducted into the asthenosphere. Rather than try to estimate how much has been extruded, however, especially given the large uncertainties in the amounts used to estimate 1,100 km, we confine our attention to the budget within the Pamir area.

BOUNDS ON THE PRESENT AND INITIAL CRUSTAL THICKNESSES

In order to evaluate the amount of continental crust subducted beneath the Pamir, we must estimate not only the amount of crust presently in this region but also the original thickness of crust. The uncertainties in each are necessarily large. Nevertheless, several observations imply both a very thick crust in the Pamir at present and a relatively thin crustal basement of the Tadjik Depression. Thus, as discussed below, the crust subducted beneath the Pamir probably was thinner than normal.

Isostasy, seismic refraction, and present crustal thicknesses

The most obvious suggestion of thick crust beneath the Pamir is the high, relatively uniform elevation exceeding 4,000 m. Crustal thicknesses beneath the other comparably high plateaus (Tibet and the Altiplano in the Andes) are approximately 70 (± 10) km (e.g., Chen and Molnar, 1981; Holt and Wallace, 1990; James, 1971; Romanowicz, 1982). The apparently small isostatic gravity anomalies over the Pamir (<20 mgal) imply local isostatic equilibrium (Artemijev and Belousov, 1983; Burov and others, 1990). Local Airy isostasy implies a thick crust.

Seismic refraction and wide-angle reflection profiles, using both explosions and earthquakes, provide the most definitive evidence for thick crust beneath the Pamir. Refracted arrivals, indicating a P-wave velocity of 8.1 km/s, were recognized only north of Karakul. The arrival times of wide-angle reflections along the long profile from north of the lake Karakul in the Pamir (Fig. 21) to the Himalaya in Pakistan imply a thickness of about 70 km, possibly reaching 75 km beneath the Northern Pamir and perhaps only 65 km beneath the Central and Southern Pamir (Beloussov and others, 1980). The suggestions of different crustal thicknesses beneath different parts of the Pamir, however, are not well established by the limited published data.

Travel-time curves corresponding to phases refracted at the Moho along profiles spanning the western edge of the Pamir and the eastern Tadjik Depression yield apparent upper-mantle P-wave velocities and intercept times that depend on the direction of propagation (Fig. 21) and hence indicate large dips of the Moho (Kulagina and others, 1974). For average crustal velocities of 6.05, 6.2, or 6.5 km/s, the reported intercept times of 13.8 s and 14.6 s at stations in the Pamir (Kulagina and others, 1974) correspond to crustal thicknesses of 63 to 79 km, with the larger average velocity and corresponding thickness more likely than the smaller ones. For paths across the eastern Tadjik Depression and the Western Pamir, from both explosions and earthquakes, Kulagina and others (1974) showed seismograms with very clear strong signals arriving later than Pn and with an apparent velocity of 6.7 km/s. Accordingly, they attributed this phase to refraction along the top of a high-velocity layer in the lower crust. They also postulated a marked low-velocity layer (V_p = 5.6 km/s beneath a layer of 6.1 km/s) within the crust of both the Pamir and the Tadjik Depression and concluded that the average crustal velocity is 6.05 km/s. In the absence of convincing evidence for the low-velocity layer, however, we assume that the average P-wave velocity is not so small and that the intercept times imply a crustal thickness beneath the Pamir of 70 km or more.

The relatively small intercept times (8.0–9.0 s) for stations in the Tadjik Depression (Fig. 21) further corroborate the existence of a marked difference in crustal thicknesses between the Pamir and the Tadjik Depression. For average crustal velocities of 6.05, 6.2, and 6.5 km/s, they imply depths to the Moho of 37 to 49 km. Because of the thick sedimentary rock within the depression, the average crustal velocity is probably lower than in the Pamir. Nevertheless, these results imply that the Moho cannot be much shallower than 40 km. Only for the southwestern part of the Tadjik Depression do seismic data suggest that this depth decreases below 35 km (to 32.5 km) (Kulagina and others, 1974).

The greater thickness of crust beneath the Pamir than beneath the Tadjik Depression is most convincingly demonstrated by the much larger apparent velocities along profiles from the Pamir westward to the depression (8.7 km/s) than in the reverse direction (7.5 km/s) (Kulagina and others, 1974). Their average suggests a Pn velocity of 8.1 km/s, and their difference implies an eastward dip of the Moho of 5 to 6°, for average crustal velocities of 6.2 or 6.5 km/s. Extrapolations of this dip along the length of the profile yield crustal thicknesses beneath the Pamir roughly 20 to 30 km greater than beneath the Tadjik Depression.

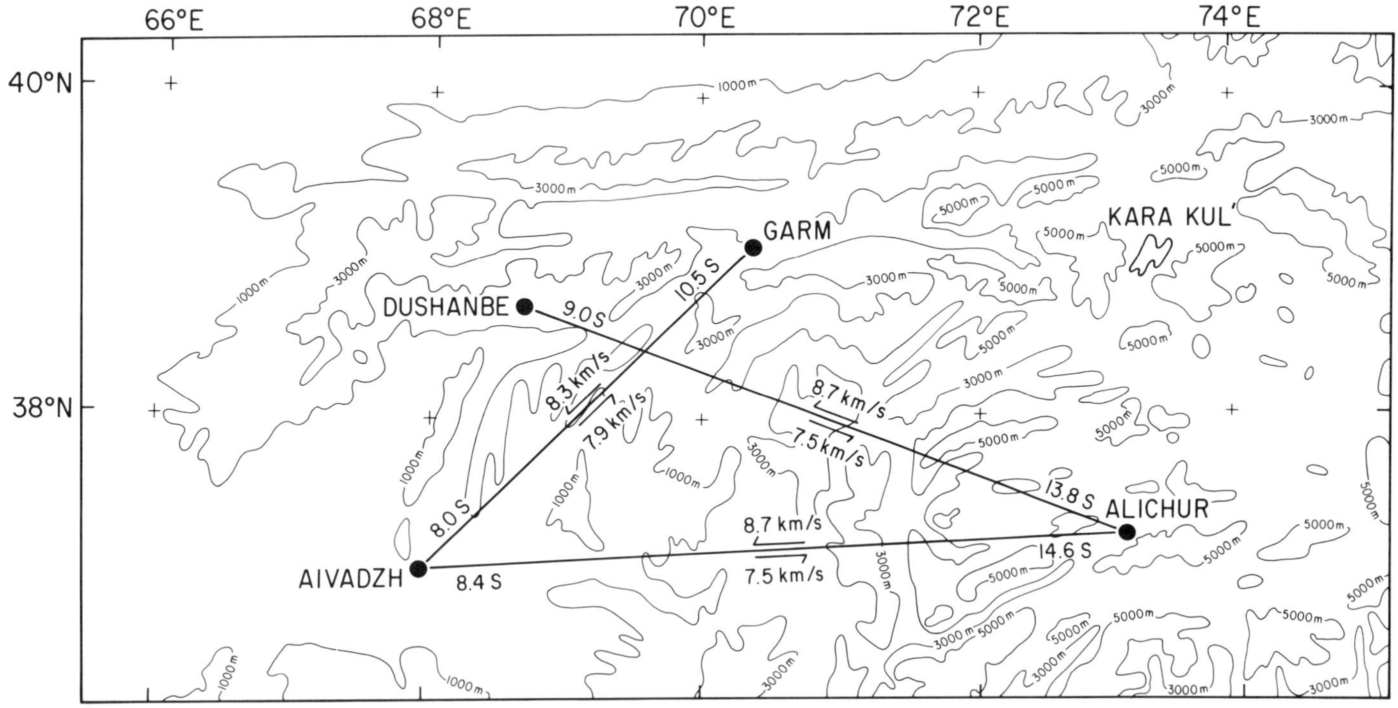

Figure 21. Map summarizing seismic refraction profiles of Kulagina and others (1974) across the western margin of the Pamir. Numbers next to lines give apparent velocities in the directions of arrows. Numbers at the ends of the lines give intercept times. Note that the apparent velocities from high to low areas are higher than in opposite directions and that intercept times at high elevations are large. These provide constraints on variations in crustal thickness beneath the region.

With very thick Mesozoic and Cenozoic sedimentary rocks in the Tadjik Depression, not only is the average P-wave velocity in the crust low but the crystalline basement must also be relatively thin. P-wave velocities in the upper part of the crust of the depression are low (4.2 and 5.3 km/s), typical of sedimentary rock. Kulagina and others (1974) deduced total thicknesses of 6 km of low-velocity material along the northern margin of the depression to 15 km in the eastern and southern parts. Although it is difficult to evaluate these thicknesses from the limited data presented, they are roughly in accord with thicknesses deduced from geologic cross sections across the Tadjik Depression (Fig. 20) (Bekker and others, 1974a, 1974b). As much as 10 to 12 km of Mesozoic and Cenozoic sediment have been deposited on a thin layer of Jurassic salt, and packages of this sequence have been overthrust atop one another. Bekker and others (1974a) reported at least 10 km and perhaps 12 km of such sedimentary rock in the western part of the depression, where the full sequence has been thrust onto the surface. They show similar thicknesses in the eastern part, but because the lower formations are not exposed, buried or blind thrust faults might make the thickness greater than the 10 to 12 km that they report.

The thick Mesozoic and Cenozoic sedimentary rock where the Moho lies roughly 40 km below the surface implies that the thickness of underlying crystalline rock is thinner than normal. In the western part of the depression, where Kulagina and others (1974) reported a depth to the Moho of 32.5 km and where the thickness of the sedimentary rock is 10 to 12 km, the thickness of the crystalline rock could be only 20 km and surely is no greater than 25 km. In the eastern part of the depression, where the depth to the Moho is roughly 40 km and the seismic data suggest a thickness of low-velocity cover of 15 km, again the thickness of the crystalline basement appears to be only 25 km and surely is less than 30 km. Given the uncertainties in the various estimates of thicknesses, it could be only 20 km. Compared with typical crustal thicknesses of 35 km or with those of 35 to 45 km of crystalline crust from the surrounding parts of Central Asia (Kosminskaya and others, 1969; Volvov'skii and Volvov'skii, 1975), the basement of the Tadjik Depression is thin. As discussed below, we presume that such a thin basement characterized the region to the east and later overthrust by the Pamir.

Subsidence of the Tadjik Depression inferred from stratigraphic sections

The stratigraphy of Mesozoic sediment suggests that thinning of the crust may have occurred prior to the deposition of the Jurassic salt. Using the measured thicknesses of the

Mesozoic-Cenozoic sequence at several localities and estimates of its compaction with time, Leith (1985) calculated the mid-Jurassic to Neogene subsidence of the basement, before terrigenous material eroded from nearby highlands was deposited. The exponential form of the subsidence history is consistent with the cooling of lithosphere thinned by 1.7 to 2.0 times (Fig. 22). If the initial crustal thickness had been 40 km or 50 km, then a thinning by 1.7 to 2.0 times would have reduced it to 20 to 25 km or 25 to 30 km. Thus, as Leith (1985) noted, the inferred thermal history and the estimates of the present thickness of crystalline basement agree with one another. Leith's (1985) inference of relatively thin crust does not apply to the entire Tadjik Depression, however, because (as discussed above) lateral variations in facies show that the center of the basin was deepest.

Leith (1985) assumed that crustal thinning occurred just before the deposition of the Jurassic salt. Alternatively, if prior to mid-Jurassic time the crust was thicker than normal, the submergence beneath sea level could have occurred well after crustal thinning had stopped. (Recall that if extension in the Basin and Range Province of the western United States stopped today, tens of millions of years would elapse before the area cooled sufficiently to drop below sea level.) The Tien Shan, much of the Tadjik Depression, and the Pamir underwent crustal shortening, and therefore presumably crustal thickening, in late Paleozoic time. If a late Paleozoic high plateau had formed, over at least part of this region, its collapse by crustal extension might have left a high, faulted landscape, similar to that present in the Basin and Range of western North America, before the area subsided below sea level. This scenario has been suggested for the Hercynian belt in central Europe (Ménard and Molnar, 1988), where Carboniferous and Permian grabens did not subside below sea level until Triassic or even Jurassic time. A similar late Paleozoic to mid-Mesozoic history of the crust between the western Hindu Kush and South Tien Shan seems plausible. Varentsov and others (1977, p. 14–20) reported Permian and Triassic conglomerate or coarse sandstone from the northern and western margins of the Tadjik Depression and inferred that such deposits characterize the region in general. Similar deposits characterize current deposition in the present Basin and Range Province and the basins of the Hercynian belt of Europe. A high plateau apparently did not extend into the Pamir, however, where Permian marine limestone was deposited. The thermally driven subsidence inferred and quantified by Leith (1985) might have begun well before the Jurassic salt was deposited.

We should note that Leith's (1985) interpretation of thermal subsidence is not the only possible explanation. Hamburger and others (1992) suggested instead that subsidence resulted from flexure of the lithosphere associated with a compressive phase of deformation in Jurassic time. The Mesozoic history of sedimentation near the Tien Shan is indeed consistent with rejuvenation of this belt in Jurassic time (Hendrix and others, 1992). Independently of the doubts of Leith's (1985) interpretation expressed by Hamburger and others (1992), Hendrix and others (1992) cautioned against assuming thermal subsidence during this period.

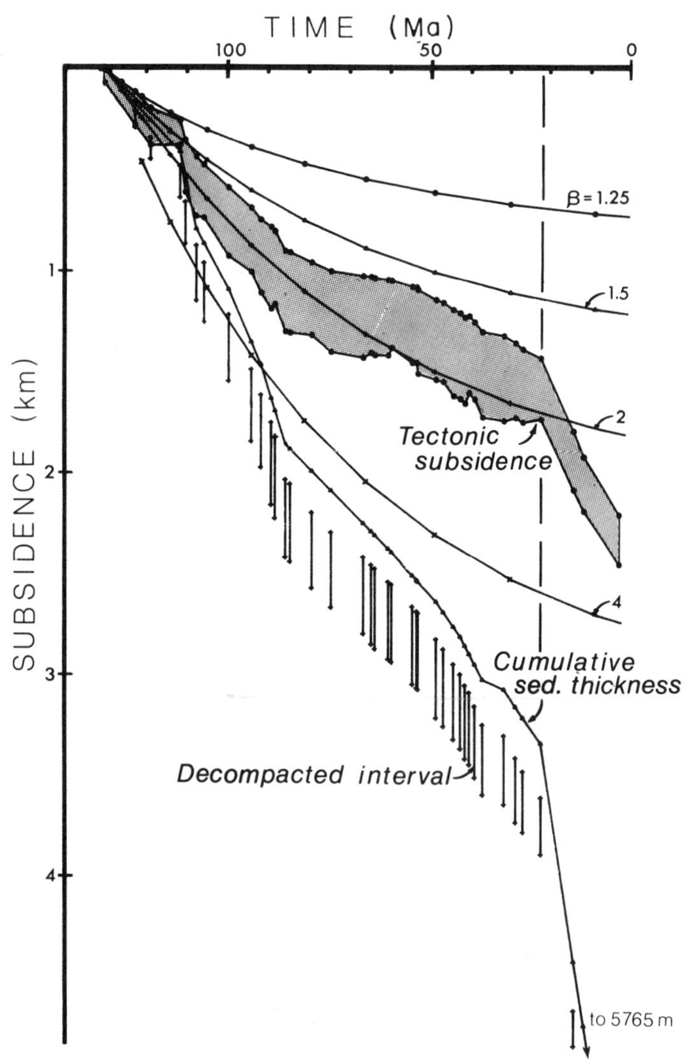

Figure 22. Plots of sediment thickness versus age with corrections for compaction and theoretical curves assuming thermally controlled subsidence (from Leith [1985]). The lowest curve shows measured sediment thicknesses for a section in the Tadjik Depression. Beneath it depth intervals show differences in corrections for various allowed estimates of compaction. The shaded region gives the range of subsidence, not due to isostatic compensation of the weight of the sediment, for different estimates of compaction. This subsidence is presumably due only to cooling and thermal contraction of the lithosphere. A family of curves for different amounts of stretching and crustal thinning shows that the subsidence history implies stretching of about 2 times and thinning to roughly half the initial lithospheric thickness.

Regardless of the physical processes responsible for it, the thickness of the early Jurassic crystalline basement of the Tadjik Depression (20–25 km) appears to be notably thinner than typical continental crust (35–40 km). Thus, the Cenozoic crustal thickening in the Pamir probably involved the southward underthrusting of relatively thin crust beneath the region that lay to the south of the eastern part of the Tadjik Depression. Accordingly, a present thickness of crust beneath the Pamir of 70 km or more cannot have been achieved simply by underthrusting of the Tadjik Depression beneath crust of normal thickness. There must have been some *in situ* shortening and thickening. The geologic structure of the Pamir, described above, demonstrates a large amount of such shortening.

ACTIVE AND QUATERNARY DEFORMATION

Several independent sources of data attest to active deformation of the Pamir, Tadjik Depression, and their surroundings, particularly along the northern margin of the Pamir. Shallow-focus seismicity occurs throughout the region but is concentrated on the margins of the Pamir (Fig. 23). Fault plane solutions of most moderate and large earthquakes ($M > 5.5$) indicate large components of thrust faulting (Fig. 24), and studies of active folding and faulting corroborate the pattern of ongoing northward overthrusting of the outer arc of the Pamir (Fig. 25). Estimates of historic, Holocene, and late Pleistocene rates of slip on faults allow rates of deformation to be estimated, and repeated geodetic measurements in the Garm region support the inference of rapid convergence across the outer arc of the Pamir.

In this section we summarize data that are relevant to the quantification of active deformation with the goal of determining how India's convergence with respect to Eurasia at about 44 (± 5) mm/a (DeMets and others, 1990) is partitioned into (1) northward underthrusting at the Himalaya, (2) distributed shortening across the Pamir, (3) southward underthrusting of southern Eurasia beneath the Pamir, and (4) crustal shortening in the South Tien Shan. These data apply to very different durations of time: 3 m.y. for plate motions, about 20 m.y. for underthrusting at the Himalaya, from a few thousand years to a few hundred thousand years for Quaternary faulting, and only a few years to a few tens of years for seismicity and geodesy. Thus, some risk is encountered by combining rates that apply

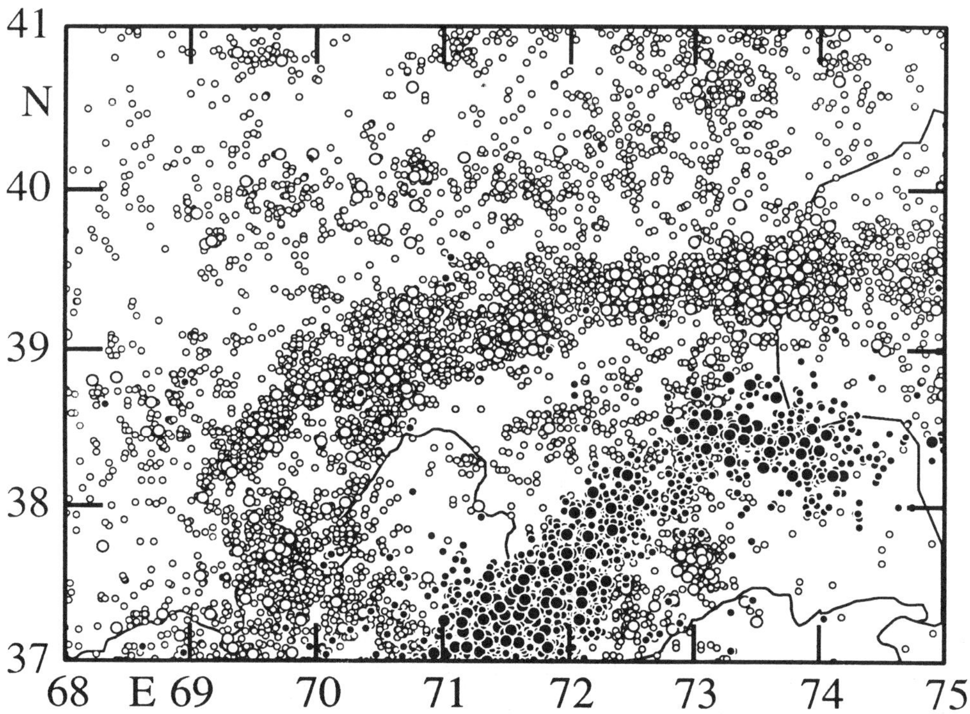

Figure 23. Map of seismicity of the Pamir and its surroundings between 1964 and 1980 located by the Central Asian regional seismic network. Open symbols indicate calculated depths less than 70 km, and closed circles indicate deeper events. Large symbols indicate events assigned local magnitudes (M_L) greater than 4.0. Note the arcuate band of shallow-focus activity along 39°N, which follows the outer margin of the Pamir, and the northeast-trending belt of intermediate-depth earthquakes farther south. The continuous line shows the borders of the former Soviet Central Asian republics with Afghanistan to the south and with China to the east. (After Hamburger and others [1992].)

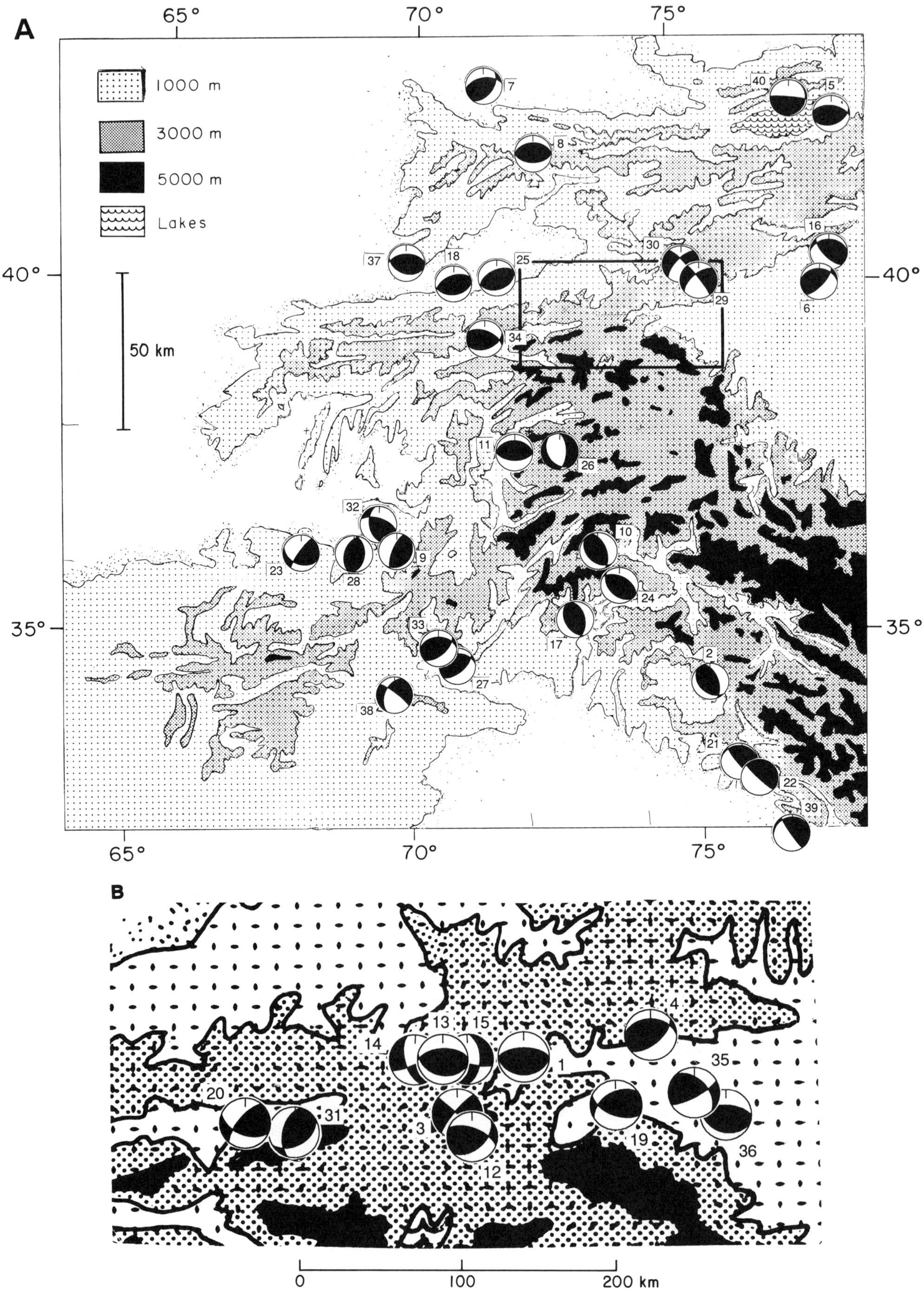
A
65°
70°
75°
1000 m
3000 m
5000 m
Lakes
50 km
40°
35°
B
0
100
200 km

to different durations, but at present there seems to be no alternative to using data that span such different intervals and assuming relatively constant tectonic processes.

Shallow-focus seismicity and fault plane solutions of earthquakes

The entire region including the Pamir, Tadjik Depression, and South Tien Shan is seismically active, but by far the most active area has been the northern margin of the Pamir (Fig. 23) (e.g., Abers and others, 1988; Gubin, 1960; Hamburger and others, 1992; Leith and Simpson, 1986; Lukk and Nersesov, 1970). The activity of the Northern Pamir is apparent also in the distribution of earthquakes since 1963 and large enough (M > 5.5) to permit the determination of fault plane solutions using seismograms from the World-Wide Standardized Seismograph Network (WWSSN) (Fig. 24). Fault plane solutions of nearly all earthquakes in this region indicate large components of thrust faulting and roughly north-south crustal shortening. Thus, this seismicity concurs with the inference that crustal shortening and crustal thickening in the Pamir are consequences of India's northward penetration into the rest of the Eurasia. Particularly large earthquakes (M > 6), however, have occurred most frequently along the South Tien Shan (Leith and Simpson, 1986). Part of India's convergence must also be absorbed along this belt.

We determined (or modified) the fault plane solutions, focal depths, and seismic moments of 13 earthquakes from the Pamir region, using a match of synthetic P and SH waveforms to those recorded by long-period instruments at stations of the WWSSN (Table 3). Appendix A provides discussions of our analysis of each of these earthquakes.

To place the seismicity of the Pamir region in the context of the surrounding area, Figure 24 shows all fault plane solutions based on data of the WWSSN for earthquakes (M > 5.5) since 1963 in the surrounding area (Table 4). Earthquakes in the western Himalaya (no.'s 2, 10, 17, 21, 22, 24, and 39 in Fig. 24 and Table 4) consistently show thrust faulting: either northeastward underthrusting of the Indian shield beneath the Himalaya or roughly northeast-southwest crustal shortening of the Himalayan crust (e.g., Molnar and Lyon-Caen, 1989). Two events southwest of the western end of the Himalaya (27 and 33) indicate a northwestward underthrusting of the Indian shield beneath the Sulaiman range, and a third (38) is consistent with a large component of left-lateral strike-slip faulting along the northeasterly trending Chaman fault in eastern Afghanistan (Ekström, 1987). Most fault plane solutions for earthquakes in the Tien Shan (5–8, 16 and 40) indicate thrust faulting and roughly north-northwest–south-southeast shortening (e.g., Nelson and others, 1987; Ni, 1978; Tapponnier and Molnar, 1979), but two (29 and 30) show large strike-slip components (Ekström, 1987) consistent with slip on the right-lateral, northwest-trending Talaso-Ferghana fault (e.g., Burtman, 1964, 1975). Largely thrust faulting and roughly north-south crustal shortening characterize the active deformation of this region.

Large components of thrust faulting characterize most earthquakes along the northern margin of the Pamir and directly to the north along the South Tien Shan (Fig. 24b) (Balakina, 1983). Assuming that the gently dipping nodal planes define the fault planes, solutions for two events along the northern edge of the South Tien Shan (18 and 25) and that of a third slightly to the north (37) suggest southward underthrusting of the Ferghana Basin beneath the South Tien Shan (Nelson and others, 1987). For most earthquakes along the northern edge of the Pamir, or along the southern edge of the South Tien Shan, both nodal planes dip at moderate angles > 30°). For these events, there is little reason for choosing which of the nodal planes is the fault plane, but the consistently nearly horizontal, nearly north-south–trending P-axes imply crustal shortening with that orientation. The large thrust components for most of these earthquakes (1, 4, 12, 13, 19, 20, 31, and 34–36) require substantial components of crustal shortening and thickening, but even those indicating largely strike-slip faulting (3, 14, and 15) also require large components of north-south shortening.

The moderate dips of the nodal planes indicate that earthquakes do not result from slip on gently dipping planes, as might be expected if the South Tien Shan were being rigidly underthrust southward beneath the northern edge of the Pamir. Instead, these fault plane solutions seem to indicate internal deformation within the upper crust along the northern margin of the Pamir. Thus, if the South Tien Shan is being underthrust beneath the Pamir, as the Quaternary faulting (discussed below) suggests, the mapped thrust faults must steepen at depth within the upper 15 km of crust where the earthquakes occur. Presumably, they flatten again at greater depth.

We can make a crude estimate of the rate of north-south shortening from the seismic moments of these earthquakes. Kostrov (1974) showed that the sum of moment tensors, $\sum M_{oij}$, in a region, divided by twice the product of the shear modulus, $\mu = 3.3 \times 10^{10} N/m^2$, times the volume of the seis-

◄

Figure 24. Maps of fault plane solutions of earthquakes that have occurred in the Pamir and its surroundings since the installation of the World-Wide Standardized Seismograph Network. Lower hemisphere diagrams of the focal spheres are plotted over epicenters, and numbers adjacent to them are keyed to Table 4. Blackened quadrants of lower hemisphere diagrams show those with compressional P-wave first motions and contain T-axes; white quadrants contain dilational P-wave first motions and P-axes. A, Earthquakes in the area surrounding the Pamir. Box in center shows area in Figure 24B. B, An enlargement of the area including the northern edge of the Pamir (the Alai and Trans-Alai ranges), where activity has been especially high in the last 30 years. Solutions 29 and 30 are not repeated in the enlargment.)

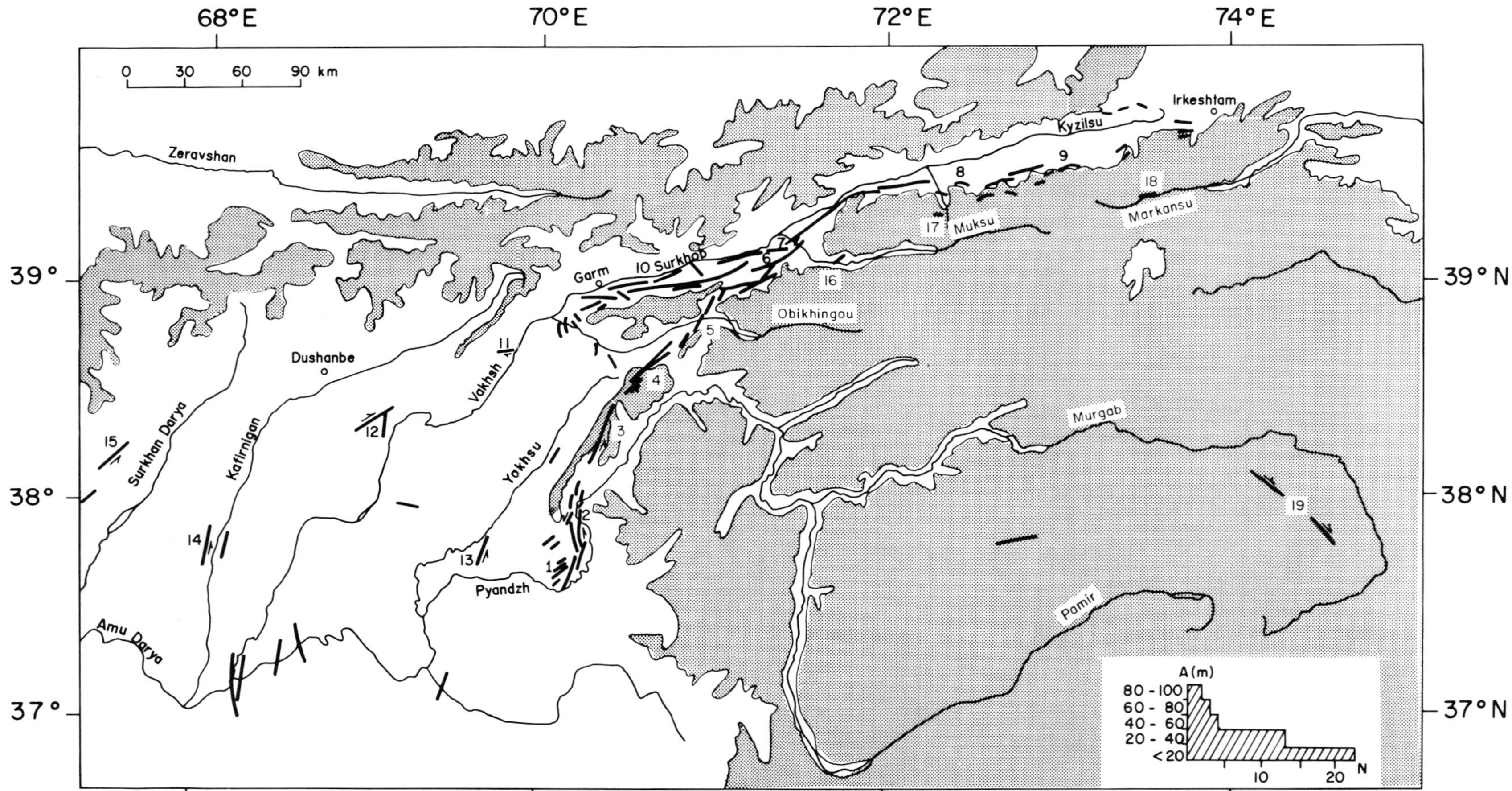

Figure 25. Map of the Pamir and the Tadjik Depression showing localities where observations of Quaternary faulting and folding were made. Areas higher than 3,500 m are shaded. Numbers correspond to localities discussed in the text. The histogram in the lower right shows the number, N, of features offset various amounts, A (m), in meters near locality no. 4 along the Darvaz fault, according to Trifonov (1983).

TABLE 3. SOURCE PARAMETERS OF EARTHQUAKES IN THE PAMIR ANALYZED FOR THIS STUDY

Date				Latitude	Longitude	Depth	Nodal Planes			M_o
Year	Month	Day	Hour	(°N)	(°E)	(km)	Strike (°)	Dip (°)	Rake (°)	(10^{17} Nm)
67	5	11	14	39.41	75.77	11 ± 3	132 ± 10	90 ± 20	169 ± 15	3*
74	8	11	01	39.38	73.81	3	291	62	131	170
						(or if two subevents)				
						6	307	87	164	89
						6	257	41	94	65
74	8	11	20	39.47	73.66	6 ± 3	267 ± 15	55 ± 5	18 ± 15	4
74	8	11	21	39.49	73.65	10 ± 3	62 +20/-15	75 +10/-8	328 +10/-15	6
74	8	27	12	39.52	73.82	7 ± 3	5 ± 15	50 +20/-15	341 +10/-12	3
78	10	8	14	39.39	74.72	18 ± 5	114 +25/-20	57 +10/-8	127 ± 10	4
78	11	1	19	39.35	72.66	5	111	52	152	98
						(or if three subevents)				
						8	116	50	180	8
						9	108	68	163	43
						10	116	50	150	44
82	9	29	04	37.35	72.94	10	0 ± 20	60 ± 10	290 ± 20	2†
83	12	16	13	39.33	72.91	11 ± 4	192 +20/-25	53 ± 10	58 ± 20	6
84	10	26	20	39.18	71.34	9 ± 3	68 +25/-20	44 +9/-6	57 ± 12	15
85	8	23	12	39.42	75.27	18	316	46	160	220
						(or if three subevents)				
						18	283	34	142	36
						19	326	58	158	79
						18	309	41	152	99
85	9	11	20	39.36	75.44	7 ± 4	108 +30/-60	31 +7/-10	104 +25/-60	8
85	10	13	15	40.32	69.84	15 ± 3	268 +15/-35	50 +7/-5	77 +12/-8	4

*Fault plane solution modified slightly from that determined from P-wave first motions by Jackson and others (1979).
†Fault plane solution modified slightly from that determined by Ekström (1987), who used a match to long-period waveforms. P-wave first motions are slightly inconsistent with his solution, but normal faulting is clear.

mogenic region, equals the *average* strain tensor for the region. The volume is given by the product of the thickness (T) of the seismogenic layer with the width (W) and length (L) of an average rectangular area that includes the seismically active region. Dividing the strain tensor by the duration (t) of the seismic record yields the average strain rate tensor. Then, the dot-product of the strain rate tensor with a vector (*W*), of length equal to the width of the zone and with an orientation parallel to that dimension, gives the velocity of motion (*v*) between the two sides of the region. The inferred velocity is independent of the width of the zone, which enters as a scalar factor both in the denominator (in the volume) and the numerator (in the width vector itself). In symbols,

$$\begin{aligned} v &= W \cdot \sum M_{oij}/(2\,\mu\,L\,W\,T\,t) \\ &= 1 \cdot \sum M_{oij}/(2\,\mu\,L\,T\,t) \end{aligned}$$

where *1* is a unit vector parallel to *W*, here chosen to point north. The sum of the seismic moments of the 11 earthquakes near the northern edge of the Pamir is

$$\Sigma M_{oij} = \begin{pmatrix} 2.1 & 0.3 & 1.2 \\ 0.3 & -4.6 & -0.3 \\ 1.2 & -0.3 & 2.5 \end{pmatrix} \times 10^{19}\ \mathrm{N\,m}$$

Using L = 400 km, W = 400 km, and T = 20 km, the average strain rate over the period of t = 25 years is

$$e_{ij} = \begin{pmatrix} 4.0 & 0.6 & 2.3 \\ 0.6 & -8.8 & -0.7 \\ 2.3 & -0.7 & 4.8 \end{pmatrix} \times 10^{-9}\ \mathrm{yr}^{-1}$$

where the 1-, 2-, and 3-axes point east, north, and up, and extensional strain rates are positive.

The strain-rate field is dominated by almost purely north-south shortening at $8.8 \times 10^{-9} \mathrm{yr}^{-1}$, which corresponds to

TABLE 4. SOURCE PARAMETERS OF EARTHQUAKES IN THE PAMIR REGION

Date			M_0*	Lat.	Long.	Depth	Nodal Planes						P-axis		T-axis		B-axis		Ref.†
Yr.	Mo.	Da.		(°N)	(°E)	(km)	Str.	Dip	Rake	Str.	Dip	Rake	Az.	Pl.	Az.	P.I	Az.	Pl.	
63	8	29	...	39.64	74.21	...	90	30	90	270	60	90	0	15	180	75	90	0	9
67	2	20	...	33.63	75.33	10	341	55	105	136	38	70	60	9	295	75	152	12	2
67	5	11	3	39.41	73.77	11	132	90	169	222	79	0	178	8	86	8	312	79	6, 10
69	9	14	...	39.64	74.87	...	270	37	124	50	60	67	156	12	276	67	62	20	9
70	6	5	31	42.48	72.71	17	69	41	68	277	53	108	354	6	242	75	86	14	8
70	7	29	...	39.88	77.80	...	270	30	136	40	70	68	147	21	278	59	48	21	9
71	5	10	1	42.85	71.29	15	37	48	57	262	51	122	330	2	235	66	60	24	8
71	10	28	...	41.86	72.37	...	270	40	90	90	50	90	180	5	0	85	90	0	9
72	6	24	19	36.28	69.69	24	43	21	109	203	70	83	298	25	102	64	205	7	1
72	9	3	2	35.94	73.33	12	341	55	105	136	38	70	60	9	295	75	152	12	2
73	10	12	...	37.68	71.88	35	90	55	90	270	35	90	180	10	0	80	90	0	3
74	8	11	170	39.38	73.81	3	291	62	131	49	48	39	353	8	252	53	89	35	10
74	8	11	4	39.47	73.66	6	267	55	81	102	36	103	3	10	145	78	272	7	10
74	8	11	6	39.49	73.65	10	62	75	328	162	59	198	18	33	114	10	220	55	10
74	8	27	3	39.52	73.82	7	5	50	341	108	76	222	335	39	231	16	123	46	10
74	9	29	...	40.39	77.98	...	270	32	49	136	66	112	209	18	80	62	306	20	9
74	12	28	14	35.06	72.91	12	354	47	102	157	44	77	75	2	334	81	165	9	7
77	1	31	11	40.11	70.86	12	81	40	100	249	51	82	344	5	114	82	253	6	5, 8
78	10	8	4	39.39	74.72	18	114	57	127	240	48	47	179	5	80	59	272	30	10
78	11	1	98	39.35	72.66	5	111	52	152	219	68	42	342	10	82	44	242	44	10
80	8	23	2	32.96	75.75	14	265	14	45	130	80	100	211	34	52	54	308	10	5, 7
80	8	23	2	32.90	75.80	13	320	5	90	140	85	90	230	40	50	50	320	0	5, 7
81	6	13	3	36.18	67.83	21	216	79	54	111	37	161	333	25	91	44	225	35	5
81	9	12	20	35.68	73.60	7	138	42	104	299	50	78	38	4	150	80	308	9	5, 7
82	5	6	5	40.15	71.54	20	70	36	95	244	54	86	336	9	138	80	246	3	5, 8
82	9	29	2	37.35	72.94	10	0	60	290	144	36	239	311	68	76	13	170	17	5, 10
82	11	20	4	34.61	70.58	5	261	33	118	48	62	73	151	15	284	69	57	15	5
82	12	16	55	36.13	68.98	22	3	43	79	225	48	100	308	3	199	82	38	7	1
83	2	13	27	39.99	75.10	15	230	60	353	323	84	211	191	25	93	16	334	59	5
83	4	5	6	40.06	75.29	9	319	74	204	222	67	342	182	28	89	5	351	61	5
83	12	16	6	39.33	72.91	11	58	47	125	192	53	58	304	3	41	65	213	25	10
83	12	22	3	36.04	69.13	15	167	48	141	286	62	49	44	8	145	53	308	35	5
84	2	1	14	34.68	70.54	4	268	38	122	50	59	69	156	11	274	68	62	19	5
84	10	26	15	39.18	71.34	9	68	44	57	290	54	118	1	6	258	67	308	35	10
85	8	23	220	39.42	75.27	18	316	46	160	60	76	46	182	18	289	42	74	43	10
85	9	11	8	39.36	75.44	7	108	31	104	272	60	82	8	15	161	74	276	7	10
85	10	13	4	40.32	69.84	15	268	50	77	108	42	105	7	4	120	79	276	10	10
86	1	12	3	34.17	69.60	9	205	45	345	306	79	226	177	39	68	22	316	43	5
86	4	26	2	32.13	76.37	13	254	16	22	143	84	105	220	37	69	49	321	15	5, 7
88	6	17	4	42.56	77.50	15	29	12	25	274	85	101	354	39	196	49	93	11	4

*Seismic moments are in units of 10^{17} Nm.

†1 = Abers and others (1988); 2 = Baranowski and others (1984); 3 = Chatelain and others (1980); 4 = Dziewonski and others (1989); 5 = Ekström (1987); 6 = Jackson and others (1979); 7 = Molnar and Lyon-Caen (1989); 8 = Nelson and others (1987); 9 = Tapponnier and Molnar (1979); 10 = new.

north-south convergence at v = 3.5 mm/a. Although the rate is relatively small, it does imply that convergence along the northern edge of the Pamir is not insignificant. More importantly, because the rate is based on only 25 years of seismicity and larger earthquakes have occurred within the area (e.g., Gubin, 1960), this rate is surely an underestimate.

The large strike-slip components for the largest of these events (12, 20, and 35) and nearly pure strike-slip faulting for others (3, 14, and 15) require that a substantial fraction of the north-south shortening be absorbed by lateral translation of material roughly perpendicular to India's convergence with Eurasia. This is apparent in the strain-rate tensor, which shows that the north-south shortening is partitioned nearly equally into crustal thickening (vertical extension) at $4.8 \times 10^{-9}a^{-1}$ and east-west extension at $4.0 \times 10^{-9}a^{-1}$. Like the north-south shortening, this east-west extension cannot be quantified reliably with only 25 years of seismic data. Nevertheless, its manifestation in the adjacent Tadjik Depression is clear. As discussed above, the north-northeasterly–trending folds and thrust faults in the depression reflect northwest-southeast shortening. Fault plane solutions of three of four earthquakes in the southern part of the depression (9, 23, and 25, but not 32) indicate a similar orientation of shortening (Abers and others, 1988). Although earthquakes with M > 5.5 have not occurred within the interior of the Tadjik Depression since the installation of the WWSSN, smaller earthquakes reveal the same northwest-southeast shortening implied by the geologic structure. Using local networks in and around the Tadjik Depression, Leith and others (1981), Lukk and Yunga (1988), and Soboleva and others (1975) determined that fault plane solutions of relatively small earthquakes (M < 4.5) indicate largely thrust faulting with northwest-southeast P-axes. Thus, although the seismicity resulting from the penetration of the Pamir into Eurasia is largely concentrated along the northern edge of the Pamir and results in relatively rapid localized crustal shortening, simultaneously this penetration seems to cause a westward translation of material in the Tadjik Depression with respect to the more stable parts of Eurasia farther north.

The fault plane solution of the only earthquake that occurred beneath the high interior of the Pamir is noteworthy in light of east-west extension of the Pamir. Although the P and S waves from the earthquake of 1982 September 29 (no. 26) recorded by long-period seismograms of the WWSSN are too small to digitize, Ekström's (1987) analysis of those recorded by digital seismographs indicates normal faulting on roughly north-south planes. Clear P-wave first motions recorded by short-period instruments of the WWSSN concur with this inference (Appendix A). Thus, this solution is similar to those of many earthquakes in the Tibetan Plateau (e.g., Molnar and Chen, 1983; Molnar and Lyon-Caen, 1989; Molnar and Tapponnier, 1978). By itself, the solution for this small earthquake does not constitute the basis for sweeping conclusions about the active tectonics of the Pamir. Normal faulting on roughly northerly trending planes, however, had been inferred by P. Tapponnier from features seen in the Landsat imagery of the eastern Pamir near the border between the Soviet Union and western China (Molnar and Tapponnier, 1978). Recent fieldwork along the eastern margin of the Pamir documents active normal faulting with a substantial east-west component (Liu and others, 1972; P. Tapponnier, personal communications, 1992). Thus, an east-west extension of the high interior of the Pamir seems to be under way. Moreover, the existence of active normal faulting suggests that crustal thickening *within* the Pamir has stopped, a conclusion that also seems to apply to Tibet (England and Houseman, 1989).

Geologic, geomorphic, and archaeologic evidence for recent deformation

Most of the observations of active faulting in the Pamir region have been made along the western and northern margins of the Pamir (Fig. 25). Recent activity follows the north-northeasterly–trending Darvaz fault, which roughly defines the western margin of the Pamir. Where this fault reaches the Obikhingou Valley (near no. 5 in Fig. 25), the Darvaz fault curves toward the east to become the Karakul fault, which separates Mesozoic and Cenozoic rocks to the north from Paleozoic rocks to the south. The zone of active faulting, however, continues northeastward and diverges from the Darvaz-Karakul fault trace. Along the Ragnov River (near no. 6 in Fig. 25), it cuts the folds in the Mesozoic and Cenozoic rocks that form the Peter the First Range. This zone of active strike-slip faulting obliquely intersects the northern margin of the Pamir, where activity follows the Trans-Alai thrust fault, along the northern slope of the Trans-Alai Range, and the Vakhsh Overthrust, along the western part of the Peter the First Range (Figs. 2 and 25). Although these belts of active faults seem to follow zones of older Cenozoic faults, in many cases the active fault planes are not reactivated older faults. Sparser traces of recent faulting have also been found within the Tadjik Depression and within the high Pamir itself. We begin with a discussion of the Darvaz-Karakul fault, continue with the Pamir-Alai zone, and then briefly consider the Tadjik Depression and the interior of the Pamir.

Active faults along the western margin of the Pamir. Most of the recent activity is associated with the north-north-easterly–trending, left-lateral Darvaz fault. Along its southern part near no. 2 on Figure 25 (in the Pyandzh Valley and the lower part of the Obi-Minyou Valley), active faults bound a north-south graben, assigned a late Quaternary age by Kuchai and Trifonov (1977) and Trifonov (1978, 1983). Evidently the graben, a few kilometers wide and roughly 20 km long, is a pull-apart basin in an overall strike-slip zone. Trifonov (1978, 1983, p. 73) assigned late Holocene ages to landforms displaced 20 m left-laterally along the strike-slip fault on the western side of the graben, early Holocene ages to terraces and alluvial fans offset 120 to 150 m, and late Pleistocene ages to landforms offset 300 m. From these he suggested a slip rate of 10 to 15 mm/a. This zone of young left-lateral displacements is particularly wide in this region; a few kilometers west of the graben, a northerly trending, en echelon system of young faults (no. 1 on Fig. 25) also attests to left-lateral displacement (Kuchai and Trifonov, 1977). This belt of active faulting continues southward into the foothills of the Hindu Kush. Farther south, Nikonov (1975) described linear depressions and tectonic scarps 50 m high, which he assigned a Holocene age.

In the upper part of the Obi-Minyou Valley (no. 3 in Fig. 25), Trifonov (1983, p. 72) reported left-lateral displacements of unspecified small landforms of a few tens of meters. Farther north, this belt of active faults curves toward the northeast. Kuchai and Trifonov (1977), Nikonov (1975, 1977), Trifonov (1978, 1983), and Zakharov (1967) described many examples of strike-slip displacements along a 20-km-long zone consisting of a series of parallel or en echelon fault

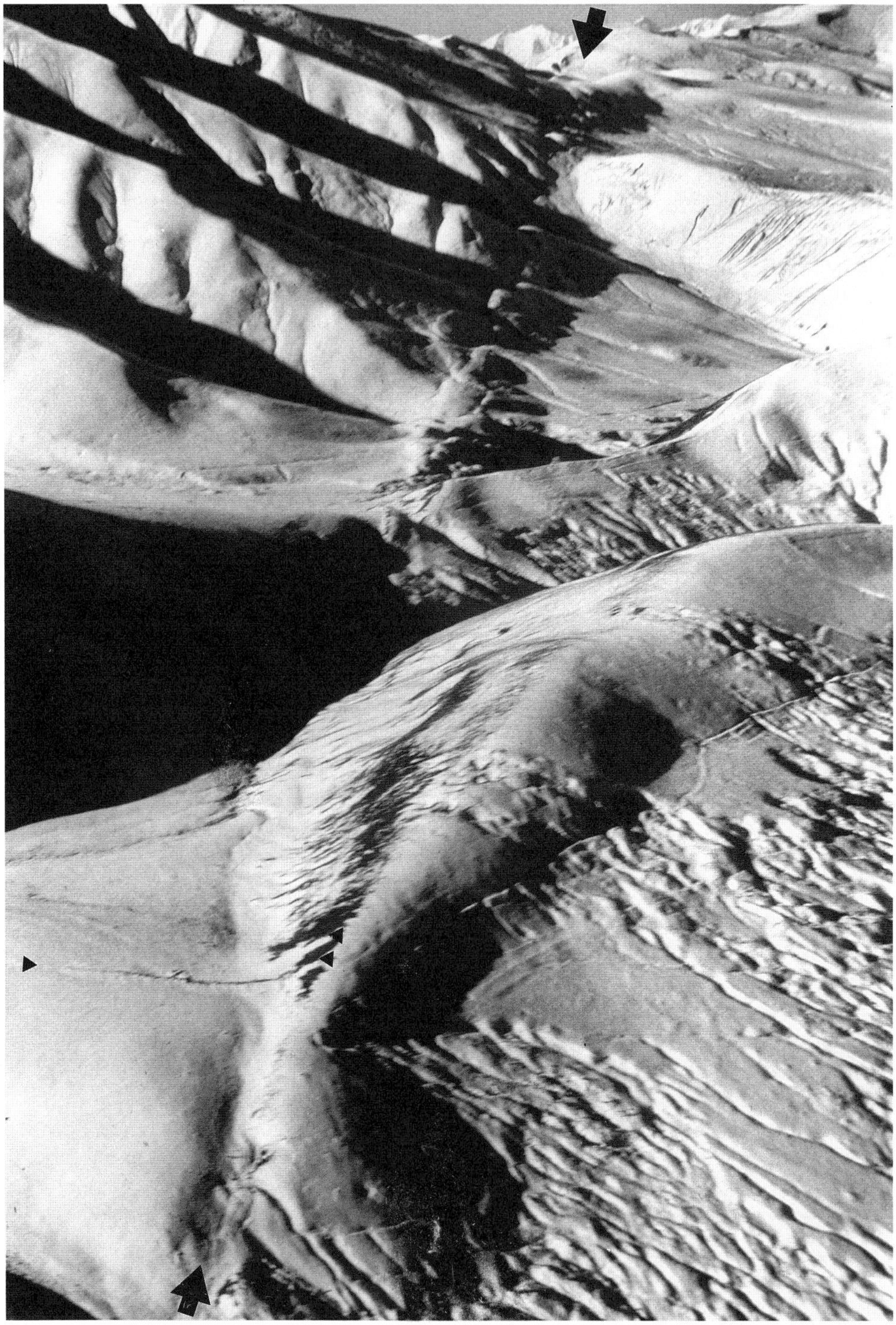

Figure 26. Photograph of the Darvaz fault looking south-southwest near locality no. 4 in Figure 25. The fault is indicated by large arrows at the top and bottom. The offset wall described by Kuchai and Trifonov [1977] and Nikonov 1983 and mapped in Figure 27 is indicated by smaller arrows on the lower left and center. (Photo by R. L. Wesson.)

strands (no. 4 in Fig. 25). The fault is especially clear here (Fig. 26). Many late Holocene dry gullies have been displaced left-laterally from a few meters to 100 m (see histogram in Fig. 25).

Among these offsets is a 21-m displacement of a man-made wall, which may have been used in the defense of nearby gold mines (Fig. 27) (Kuchai and Trifonov, 1977). Nikonov (1977, p. 127) reported that gold mining ended in the fourteenth or fifteenth century, but Kuchai and Trifonov (1977) associated the mining with the Sogdian culture and inferred an age of 1,500 years. Because Sogdiana was a well-defined political entity at the time of Alexander the Great, however, an age of 2,200 years for the wall is possible. In any case, this offset implies that the rate of slip has been many millimeters per year (between roughly 10 and 40 mm/a).

Kuchai and Trifonov (1977; Trifonov, 1978, 1983, p. 71]) assigned 160-m offsets of dry valleys in the same region to the entire Holocene epoch and 800-m offsets of two valleys to the interval spanned by the late Pleistocene and Holocene epochs (~70,000 years). In addition, Nikonov (1975) reported an offset of 3 km of conglomerates from the Kilimbin formation. He stated that this formation was deposited in the interval between 2.5 and 0.7 Ma, suggesting that the offset occurred since 0.7 Ma. These amounts of offsets and their assigned ages indicate average rates of strike-slip displacement along this part of the Darvaz fault of 5 to 16 mm/a for the Holocene epoch and 4 to 12 mm/a since the beginning of the late Pleistocene epoch (Nikonov, 1975, 1977, p. 126; Trifonov, 1983, p. 72), with Nikonov favoring the lower part of each range and Trifonov the higher part. Because the ages of all landforms and deposits are based on correlations with features and deposits in the Tadjik Depression and because no quantitatively measured ages were reported for samples from the vicinity of the Darvaz fault, the rates of slip may be more uncertain than Nikonov (1975, 1977) and Trifonov (1983) suggested.

Farther northeast, Trifonov (1983, p. 66) reported Holo-

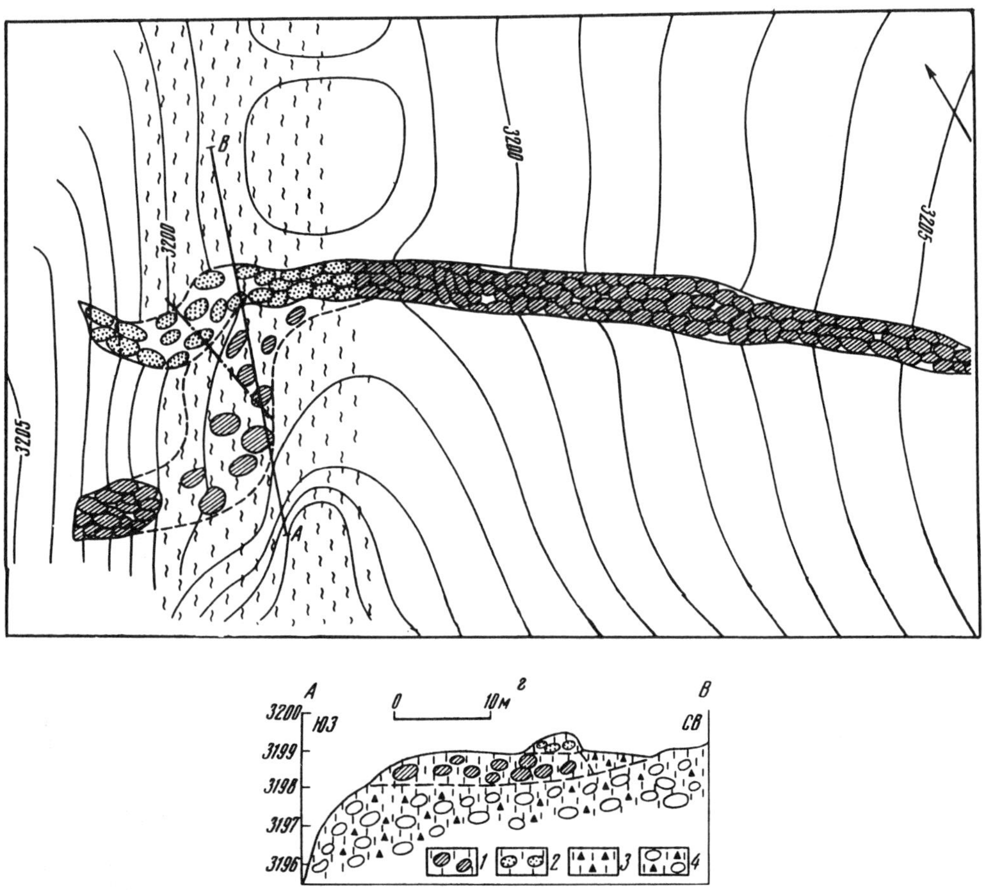

Figure 27. Detailed topographic map of the region near the offset Sogdian wall and cross section across it near locality no. 4 in Figure 25. As can be seen in the photo in Figure 26, the fault zone (squiggles on left side of map) is relatively wide (about 20 m). The wall is bent at the fault. The initial wall, indicated by stones with dark, diagonal hatching, has been offset 21 m. After a fraction of this offset occurred, a newer segment, indicated by stones with dots in them, was built and has been offset about 5 m. Contours are in meters above sea level. Arrow in upper right points northward. As shown in cross section A-B, the walls were built on Quaternary alluvium and colluvium, indicated by open stones and triangles. (After Kuchai and others, 1981; Kuchai, 1983.)

cene displacements of up to 10 m of small valleys along the southern branch of the Obikingou Valley (no. 5 in Fig. 25). At the Karashura Valley (no. 6 in Fig. 25), Trifonov (1983, p. 66) described a 170-m, left-lateral displacement of a glacial valley and the moraine left in it, which he assigned a late Pleistocene age. If this moraine formed since the last glacial maximum, since about 20 ka, then the slip rate would be 8 mm/a or more. Farther northeast, the fault zone splays into a wide zone. Near where the Muksu Valley reaches the Surkhob (no. 7 in Fig. 25), both the margins and small ridges (a few meters high) along the surface of a lateral moraine have been displaced left-laterally 50 m, and the south side has been uplifted 10 m relative to the north side (Fig. 28). Trifonov (1983) also associated this moraine with the last glaciation. As it is cut by only one splay, the offset of the moraine gives only a lower bound of 3 to 4 mm/a on the total late Pleistocene offset.

In summary, the offset features with various ages collectively support the inference that the slip rate is 10 to 15 mm/a on the Darvaz fault. Little evidence has been presented to support the ages assigned to most of the land forms displaced along the Darvaz fault, however, and the credibility of this rate relies largely on three of the offset features: the Sogdian wall (Fig. 27), the valley containing a late Pleistocene moraine (no. 6 in Fig. 25), and the offset lateral moraines along one of several splays of the fault zone (Fig. 28).

The Trans-Alai belt of active faulting. The zone of strike-slip faulting that follows the Darvaz fault seems to end where it intersects the Trans-Alai Range, near where the Kyzilsu and Muksu rivers join (Fig. 25). Active deformation, resulting largely from thrust faulting, follows the northern margin of the Pamir along the northern slope of the western Peter the First and the Trans-Alai ranges, where linear furrows and tectonic scarps can be traced along nearly the entire range front. Nikonov and others (1983) noted that numerous landslides, rockfalls, and avalanches, as well as tension cracks, characterize this zone of young tectonic deformation. They associated these features with preinstrumental earthquakes. Carbon-14 ages and investigations of lichens from many of these features indicate Holocene ages (near no. 8 and no. 9 in Fig. 25). More importantly, Nikonov and others (1983) presented abundant evidence for Quaternary folding, thrust faulting, and tilting of Quaternary terraces along the Trans-Alai range (Figs. 29, 30, 31). For example, along the left bank of the Surkhob just below the confluence with the Muksu, they showed early Cretaceous rock thrust some 500 m over flat-layered Quaternary cobbles, gravel, and sand (Fig. 29) (Nikonov and others, 1983, p. 83). Unfortunately, they could not date the young sediment precisely enough to determine a slip rate.

In the vicinity of locality no. 8 (Fig. 25), Nikonov and others (1983, p. 115) described the folding of middle Pliocene conglomerate and the overthrusting of middle Pleistocene conglomerate onto very young river gravels on a fault dipping south-southwest (Figs. 30 and 31). In one locality, late Pleistocene fluvioglacial gravel has been thrust 5 m onto soil (Fig. 31).

Figure 28. Photograph of offset lateral moraines near the Muksu River. View is east-southeast, looking upstream along the Muksu River near localtiy no. 7 in Figure 25. The fault trace enters the photo in the lower right and crosses the lower middle of it. The southern, far side of the fault appears to be displaced left-laterally and upward relative to the northern, near side. According to Trifonov (1983), this trace is only one of a number of splays from the Darvaz-Karakul fault. (Photo by V. G. Trifonov.)

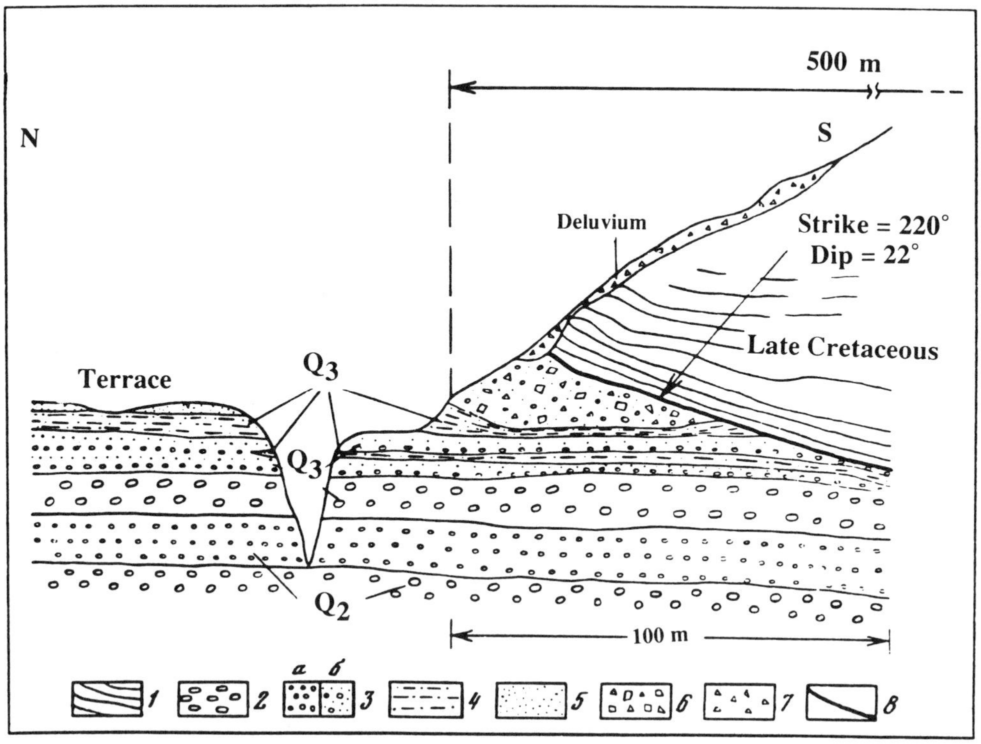

Figure 29. Cross section across the frontal part of the Vakhsh Overthrust near the confluence of the Surkhob and Muksu rivers (Fig. 25). Highly fractured late Cretaceous rock has been thrust over middle (Q_2) and late Quaternary gravel (Q_3). Legend: 1—fractured late Cretaceous rock; 2—cobbles; 3—gravel, a—pebbly, b—gritstone; 4—sandy loam; 5—sand; 6—fragmented debris (colluvium); 7—sandy soil mixed with debris; and 8—thrust surface. (Modified slightly from Nikonov and others [1983].)

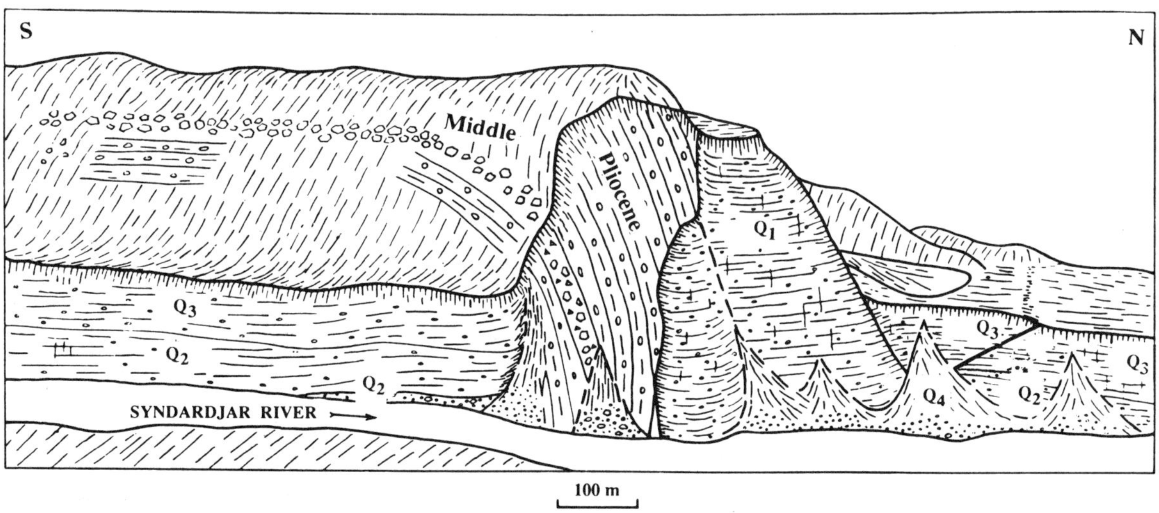

Figure 30. Sketch of folded middle Pliocene conglomerate and thrust fault (heavy line) cutting middle Pleistocene (Q_2) fluvioglacial conglomerate and late Pleistocene gravel (Q_3), near locality no. 8 in Figure 25. Holocene colluvium (Q_4) has slumped down into the river valley to cover some of the fault. (Modified slightly from Fig. 6 of Nikonov and others [1983].)

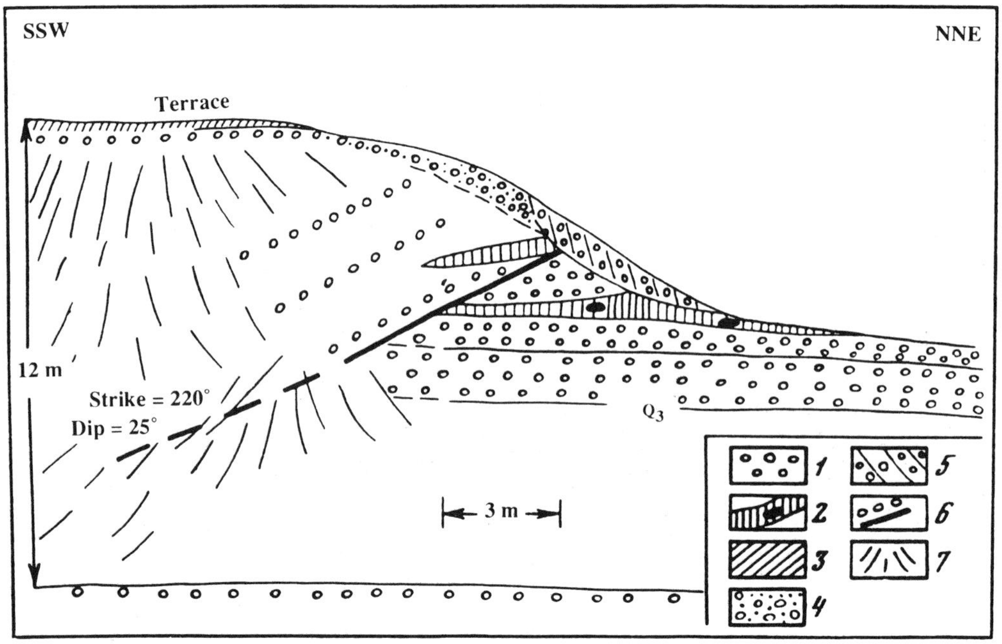

Figure 31. Cross section across active thrust fault, near locality no. 8 in Figure 25. Terrace gravel has been thrust about 5 m over late Pleistocene gravel (Q_3) and young soil. Legend: 1—late Pleistocene fluvioglacial terrace gravel, 2—buried soil with charcoal (no date was given), 3—modern soil, 4—early generation of colluvium, 5—late generation of colluvium, 6—fault surface, and 7—talus. (Modified slightly from Nikonov and others [1983].)

In a nearby valley, below a horizontal layer of river gravel 5 m thick, Nikonov and others (1983, p. 115) found buried remnants of a Stone Age encampment, now consisting of a layer 0.25 m thick and 5 m in horizontal extent. The stone artifacts (weapons, plates, and scrapers) imply a Neolithic age, which is confirmed by a Carbon-14 age of 5,100 ± 150 years, or between 5.75 and 6.00 ka when corrected for variations in cosmic-ray production of ^{14}C (Pearson and others, 1986). Nikonov and others (1983) measured a horizontal component of slip of middle Pleistocene conglomerate over these flat Holocene deposits of 12 to 15 m (and suggested the possibility of 25 m). These facts suggest a minimum rate of thrust slip on this fault since 6.0 ka of 2.0 to 2.5 mm/a (and possibly 4.0 mm/a).

Nikonov and others (1983, pp. 162–173) described numerous examples of macroseismic effects (small scarps, tension cracks, landslides, rockfalls, ice falls, avalanches, etc.) associated with the Alai earthquake of 1978 November 1 (M = 6.8) (no. 20 in Fig. 24), whose epicenter lies near locality no. 8 in Figure 25. Most of these examples lie within a zone some 20 km long along the Trans-Alai thrust fault, which separates the Trans-Alai Range from Quaternary deposits in the Alai Valley. Virtually all orientations of scarps and tension cracks are represented, but many more are aligned east-west than north-south. Nikonov and others (1983, p. 166) reported one example of a west-northwest–trending zone of cracks and scarps, some 1.5 km long, with right-lateral displacements of 100 to 150 mm and vertical components of 100 to 200 mm with the south side up. Even this example, however, does not seem to represent primary surface faulting. In any case, Nikonov and others (1983) related these ruptures to adjustments within the upper block of Trans-Alai thrust fault, an inference corroborated well by the fault plane solution of this earthquake (Figs. 24 and A6) (Balakina, 1983; Ekström, 1987).

Nikonov and others (1983, pp. 142–144) also described macroseismic effects of the 1974 August 11 Markansu Valley earthquake (M = 7.4) and its aftershocks (no's. 12–15 in Fig. 24), which occurred in the eastern part of the Trans-Alai region. The instrumental epicenter suggests that the mainshock occurred on the southern slope of the Trans-Alai Range along the Markansu Valley, but a search for surface deformation in 1975 revealed no significant traces of it there. In 1979, Nikonov and others mapped landslides and avalanches along the north slope of the Trans-Alai Range, which resembled those clearly associated with the earthquake of 1978 November 1. Moreover, in 1975 Nikonov found east-west–trending tensions cracks in surface deposits in glaciers along the north slope of the Trans-Alai Range, both of which Nikonov and others (1983) confidently assigned to the earthquake in 1974. In only one locality did Nikonov find what might be evidence of surface faulting: a steeply south-dipping crack between two layers with possibly 100 to 300 mm of displacement. Clearly the earthquake did not produce a major surface rupture.

Active faulting along the north slope of the western Peter the First Range. The western part of the Peter the First Range forms a western continuation of the Trans-Alai Range, and the deformation found along the foot of the Trans-Alai continues westward as a wide belt, some 9 to 12 km in width, on the northern slope of the Peter the First Range and along the Surkhob valley. The Surkhob belt of active faults continues west of the Muksu river as a belt of active thrust and reverse faults and diverges from the northeasterly trending Darvaz belt of largely strike-slip displacement, described above. Linear furrows and scarps, similar to those along the Trans-Alai belt, are common. Carbon-14 ages of 670 ± 40, 1,470 ± 100 and 2,000 ± 100 years obtained by Trifonov (1983) from swamp deposits (no. 10 in Fig. 25) that accumulated in basins dammed by tectonic scarps suggest recent tectonic activity. Trifonov (1983, p. 83) reported the sharp right-lateral bending of a dry valley of 10 to 15 m and other much smaller right-lateral offsets along west-northwesterly faults from this same locality. The most definitive evidence of active deformation of this area, however, is provided by geodetic measurements, discussed below.

Active faults in the Tadjik Depression. Zakharov (1948, 1955, 1958, pp. 183–200) presented a number of arguments for Quaternary right-lateral displacements along the major faults on the north edge of the Tadjik Depression, and Nikonov (1974, 1977, Fig. 29) related Holocene scarps, linear furrows, and landslides to earthquakes along the foothills of the Gissar and Karategin ranges on the northern border of the Tadjik Depression. Legler and Przhiyalgovskaya (1979) described displacements of 15 m of Holocene landforms along two parallel, east-west–striking, right-lateral faults (no. 11 in Fig. 25) that follow the foot of the Karategin Range. South of Dushanbe, late Pleistocene terraces have apparently been displaced 90 m right-laterally along the northeasterly trending Yavan fault (no. 12 in Fig. 25) (Trifonov, 1978, 1983, p. 87).

Nikonov (1970) and Trifonov (1983, pp. 88–89) reported evidence of the late Quaternary displacements along some of the oblique-reverse faults within the Tadjik Depression: 40 to 55 m of left-lateral offset of unspecified late Pleistocene landforms along a north-northeast–trending fault east of the Yaksu River (no. 13 in Fig. 25), Holocene left-lateral strike-slip offsets of 2 to 2.5 m of small dry valleys in the Kafirnagan Valley (no. 14 in Fig. 25), and left-lateral displacements of 10 to 35 m of watersheds and dry valleys west of the Surkhan Darya since late Pleistocene time (no. 15 in Fig. 25). These observations collectively imply active northwest-southeast shortening of the Tadjik Depression. Despite the uncertainties in ages, the small number of observations of recent faulting and the small displacements suggest that the interior of the Tadjik Depression deforms much less rapidly than the margins of the Pamir.

Active faults within the high Pamir. Several observations suggest active deformation in the northern extremity of the Pamir. Nikonov and others (1983, p. 105) reported young linear depressions and scarps along the Karakul fault in the Muksu Valley in the Northern Pamir (no. 16 in Fig. 25). In one place a fault plane is seen to dip south at 50 to 70°. Landslides, inferred to be consequences of young seismic activity, are common eastward along this zone. In the Altyn Dara Valley (no. 17 in Fig. 25) there are scarps 20 to 40 m high. A defensive wall built in the Middle Ages across one of the scarps along the Altyn Dara shows no evidence of deformation, suggesting that that scarp formed more than about 1,000 years ago. Nikonov and others (1983, p. 107) also reported linear depressions, 20 to 30 m deep, and Pleistocene landslides, possibly of seismic origin, along the eastern continuation of this fault zone in the Markansu and Uisu Valleys (no. 18 in Fig. 25). Thus, the deformation in the northern Pamir is not confined to the north slope of the Trans-Alai Range but forms a belt some 20 km wide.

Within the interior of the Pamir, Trifonov (1983, p. 74) reported that Pleistocene river terraces with different ages have been displaced horizontally 35 to 70 m by the Karasu strike-slip fault (no. 19 in Fig. 25). This fault is the western strand of the Aksu-Murgab fault zone with roughly 80 km of Cenozoic right-lateral displacement (Ruzhentsev, 1963, 1968, pp. 119–143). In addition, Nikonov and others (1983, p. 98) noted landslides at localities along the main east-west valleys in the Pamir, which they associated with young tectonic activity. In any case, evidence of active faulting seems to be much less abundant within the Pamir than along its western and northern margins.

Summary of Quaternary and Holocene faulting. The most rapid deformation seems to be concentrated on the margins of the Pamir: along the left-lateral Darvaz strike-slip fault, by folding and thrust faulting at the Trans-Alai Range, and possibly also by oblique thrust faulting on the northeast margin of the Pamir, as suggested by the fault plane solutions of recent earthquakes (Fig. 24). The Pamir seems to be moving northward, past the Tadjik Depression to the west, at roughly 10 to 15 mm/a along the Darvaz fault. Crustal shortening by folding and thrust faulting and possible modest counterclockwise rotations within the Tadjik Depression require that the Pamir move northward with respect to the South Tien Shan at a somewhat higher rate. As discussed below, we suspect that north-south crustal shortening across the Trans-Alai Range could reach 20 mm/a. The apparently low level of activity *within* the Pamir suggests that the entire high range moves northward toward the South Tien Shan at roughly this rate. If the 10 to 15 mm/a rate of strike slip on the northeasterly trending segment of the Darvaz-Karakul fault and the rate of convergence between the Pamir and the South Tien Shan of about 20 mm/a are reliable, then the north-south convergence rate across the Garm region should be only 10 to 15 mm/a, west of where the Darvaz strike-slip system intersects the Trans-Alai Range (Fig. 25). As discussed below, geodetic results from the Garm region offer support for this suggestion.

Geodetic evidence of rates of deformation in the Garm Region

The Garm region spans the western part of the Peter the First Range, which consists of folded and overthrust Mesozoic and Cenozoic sedimentary rocks of the northeast corner of the Tadjik Depression (Fig. 17). In the southern part of the region, the Karakul fault separates Paleozoic rock of the Pamir to the south from the Cretaceous and Cenozoic sedimentary rock, which has been detached from the underlying rock along a layer of late Jurassic evaporites (e.g., Hamburger and others, 1992). These Cretaceous and Cenozoic rocks crop out in folds and imbricate thrust sheets in the Peter the First Range. The entire package of Mesozoic and Cenozoic rock has been thrust northward onto Paleozoic metamorphic rock of the South Tien Shan at the Vakhsh Overthrust (Figs. 17, 19, and 20).

As part of a detailed investigation of earthquakes and active deformation, geodetic measurements of many types have been made, primarily across or near the Vakhsh Overthrust along the northern edge of the Peter the First Range, but also to determine regional strain across the whole region. The Vakhsh Overthrust has been investigated in two adjacent regions: a triangulation and leveling network, some 6 km in length, along the Runou valley, and a comprehensive investigation of deformation near the active scarp just south of Garm at Sari-Pul' (Fig. 32). In addition, a network of benchmarks installed on peaks and high areas of the Peter the First Range and its surroundings was surveyed with theodolites in 1966–1967 and in 1974 and then with laser distance-measuring devices beginning in 1974. We discuss the network in the Runou Valley first, then the area near Sari-Pul', and last the regional network.

Runou Valley. in 1948–1950, 21 benchmarks along a line trending northwest-southeast from the crystalline rock of the South Tien Shan southeastward up the Runou Valley (Fig. 32) were installed and surveyed, and they were resurveyed in 1968 and later (Guseva, 1986; Konopaltsev, 1971). Guseva (1986, p. 66) reported fractional errors in baseline distances of 1.4 mm/km for the earlier surveys and 1 mm/km for the more recent ones and corresponding root-mean-square (rms) errors in angles of 1.0" and 0.6", or 5×10^{-6} and 3×10^{-6}. Thus, rms errors in positions of points 200 m apart, other than those of baselines, would be about 1 mm. Those between the extremities of the network—some 6 km apart—could reach 20 to 30 mm.

Guseva (1986, p. 151) and Konopaltsev (1971) reported no resolvable movements between benchmarks on the crys-

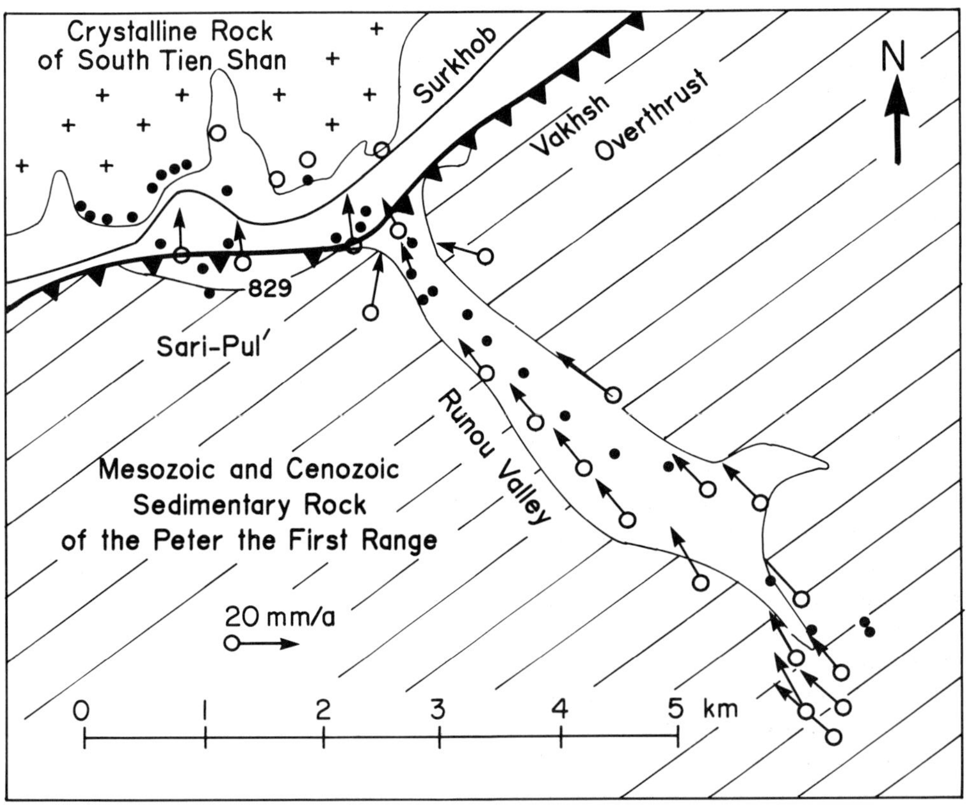

Figure 32. Map of the region near Garm, on the northern edge of the Peter the First Range, showing geodetic benchmarks along the Runou Valley and near Sari-Pul'. The town of Garm (see Fig. 25) lies 1 km north of the middle top edge of the map. Open circles show locations of points used in triangulation and trilateration measurements, and black dots show points used to infer vertical displacements. Arrows show the results of triangulation in 1984–1949 and retriangulation in 1968 along the Runou Valley and between 1972 and 1980 near Sari-Pul', in both cases with respect to points north of the Surkhob River. Benchmark no. 829 lies near a tunnel cut into the Vakhsh Overthrust (see Fig. 34), where repeated measurements have been made.

talline rock of the South Tien Shan, but between 1949–1950 and 1968 (18.5 years), nearly all points south of the trace of the Vakhsh Overthrust were displaced north-northwest (302 to 323°) at average rates of 13 to 23 mm/a (32 and 33) (or a total of about 240 to 420 mm). Guseva (1986, p. 151) reported an average velocity of 15.8 mm/a at 317°. She listed uncertainties of 1 mm/a (or less) and 2 to 5° for velocities of points relative to the South Tien Shan, which seem a little small if uncertainties in relative positions reached 20 to 30 mm for each survey. In any case, the data certainly permit, if they do not require, some relative movement among benchmarks within the Peter the First Range. Most of the convergence across the region between 1949–1950 and 1968, however, apparently was localized at the Vakhsh Overthrust, where the average rate was roughly 16 mm/a (Fig. 33).

Subsequent remeasurements, using annual retriangulations of the network from 1968 to 1972 and then beginning in 1972 using laser-ranging devices to measure directly distances between some of the benchmarks, revealed a more complicated pattern. Measurements of distances along lines do not change linearly with time, nor does the measured strain appear to be as concentrated near the Vakhsh Overthrust as for the earlier period (Guseva, 1986, pp. 68–69, 152). Guseva's (1986) results seem to suggest distributed compressive strain across the region. If, however, the uncertainties in relative positions in each survey was as large as 20 to 30 mm, then 2 to 4 years would be too short to resolve rates of 10 to 20 mm/a. Hence, the more complex deformation indicated by surveys in 1968–1972 than between 1949–1950 and 1968 might be due to noise. Guseva (1986, p. 69) also reported a 10-year average rate of shortening of 11.6 mm/a (1974–1984) of a 5-km-long line that spans the Vakhsh Overthrust but of 3.9 mm/a of a 4-km-long line that does not cross it. These results suggest only 8 mm/a of convergence across the overthrust, but 4 mm/a of shortening within the northernmost 4 km of the Peter the First Range. Moreover, she reported an average rate of convergence of 10.5 mm/a between 1972 and 1975 between two points at the south end of the Runou Valley. As little was written about uncertainties, it is difficult to know if these more recent data suggesting nearly 15 mm/a of compressive strain *within* the Peter the First Range are inconsistent with those of Konopaltsev's (1971) earlier surveys showing a comparable amount localized at the Vakhsh Overthrust.

Repeated leveling in the 1970s of the benchmarks installed in 1948–1950 and in 1969–1972 also suggest a localization of deformation near the Vakhsh Overthrust (Figs. 32 and 33). Although the points surveyed only since 1969 show more scatter, they reveal a pattern similar to those surveyed initially in 1948–1950. Little relative vertical movement among benchmarks northwest of the fault contrasts with average rates of several millimeters per year (over durations of 5 to 29 years) of benchmarks southeast of the fault (Fig. 33) (Guseva, 1986, p. 149). The difference in vertical speeds near the fault is 7 to 9 mm/a but decreases abruptly to 1 to 4 mm/a

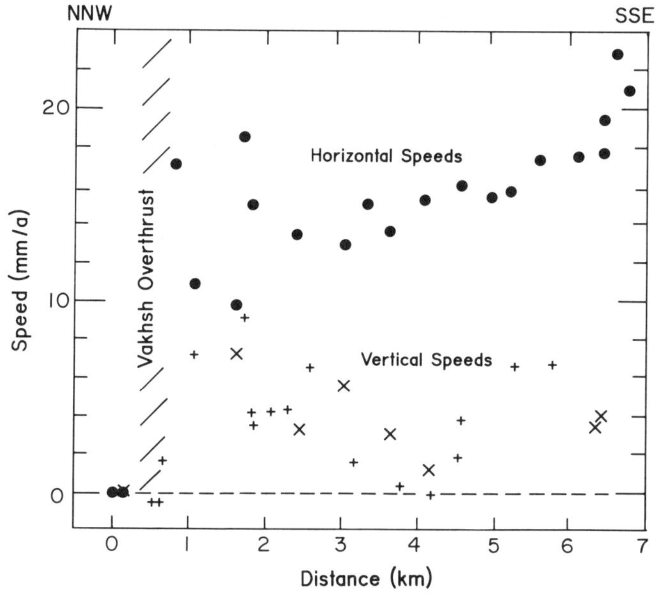

Figure 33. Profile of horizontal and vertical speeds of benchmarks along the Runou Valley (Fig. 32). Black dots show average horizontal speeds between 1948–1949 and 1968, according to Konopaltsev (1971). Crosses show vertical speeds averaged over durations of 25 to 30 years (from 1948 to 1978), and smaller plus signs show vertical speeds averaged over only 9 years or less (between 1969 and 1978). (Taken from Guseva [1986].)

toward the southeast. This suggests that if the vertical movement is due to slip on a single southerly dipping thrust fault, the fault is listric, and the dip of the fault decreases rapidly with depth southward.

Sari-Pul' network. In 1972, a network of benchmarks was installed on opposite sides of the Surkhob river valley at Sari-Paul', near Garm, where the Vakhsh Overthrust manifests itself as a prominent scarp. Repeated triangulation of the network yielded distances between benchmarks on opposite sides of the scarp that decreased steadily, if not linearly, with time. Guseva (1986, p. 73) reported average northward velocities of the two points on the hanging wall of the Vakhsh Overthrust of 15.9 ± 0.5 mm/a toward 353° ± 2° and 15.8 ± 0.6 mm/a toward 351° ± 2°, with respect to the points on the South Tien Shan, for the period 1972–1980.

Remeasurements of vertical angles between points on opposite sides of the scarp revealed rapid uplift at about 9 and 15 mm/a of the two points on the south side with respect to those in the South Tien Shan. Thus, relative to the South Tien Shan, measured speeds of the points are 18.6 and 21.6 mm/a, at angles of 31° and 43° with respect to the horizontal (Guseva, 1986, p. 73). The relative magnitudes of the vertical speeds of these two benchmarks, some 400 m apart, and that at the mouth of the Runou Valley 1 km farther east are consistent with the scarp being most prominent near point 829 and least so at the Runou Valley. If these speeds have been accurately measured, however, the differences among them imply that the Vakhsh Overthrust is not a simple planar fault separating two essentially rigid blocks.

In the early 1970s, a tunnel was dug across the Vakhsh Overthrust at Sari-Pul', and strainmeters, tiltmeters, and closely spaced benchmarks (4 to 4.5 m apart) were installed. Elevation changes measured using both classical leveling techniques and water-tube tiltmeters showed a steady rise, at nearly a constant rate, of the southern end of the tunnel with respect to the northern end. The releveling yielded a rate of 14.6 ± 0.3 mm/a between 1973 and 1976, with a smooth variation in rate along the tunnel (Fig. 34) (Guseva, 1986, p. 51). The rapid convergence and uplift suggest that the Vakhsh Overthrust dips moderately steeply southward. Drilling into the hanging wall 113 m south of the scarp corroborated such a dip. The borehole penetrated some 20 m of Quaternary alluvial gravel, 123 m of Cretaceous red beds, and then Quaternary alluvial deposits again at a depth of 143 m (Fig. 34) (Guseva, 1986, p. 101; Kuchai and others, 1978). The average dip of 48° accords reasonably well with the inclinations of the slip vector (43°) deduced from the vertical rate of 14.6 ± 0.1 mm/a (for 1966–1983) (Guseva, 1986, p. 142) and the horizontal rate of 15.8 ± 0.6 mm/a (for 1972–1980) of benchmark 829, which lies just south of the south end of the tunnel.

Regional triangulation and trilateration network. The results from the Sari-Pul' region suggest rapid overthrusting of the Peter the First Range onto the South Tien Shan at the Vakhsh Overthrust. The data of Konopaltsev (1971) from the Runou Valley can be taken as evidence that the Peter the First Range moves northward with little internal deformation. A key question is whether this movement represents a part of India's penetration into Eurasia, or whether it could be superficial deformation, such as that resulting from a major listric slump (or

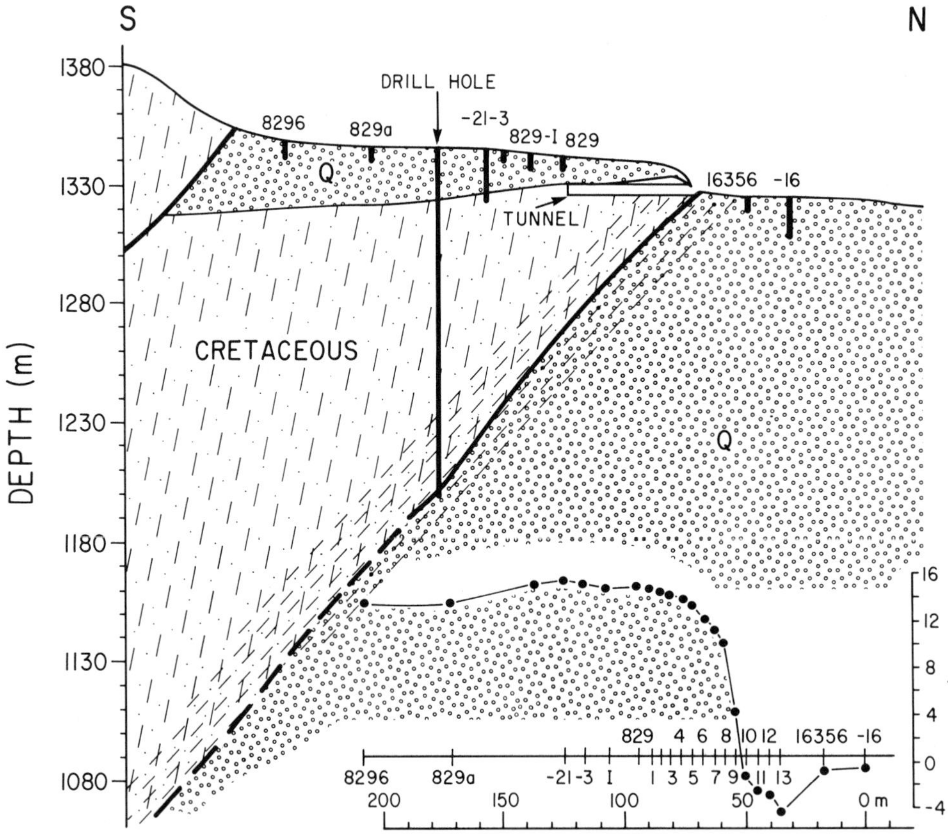

Figure 34. Cross section of the northern edge of the Peter the First Range near Sari-Pul' (near benchmark no. 829 in Fig. 32), showing the locations of a drill hole through the section, a tunnel dug at the fault scarp, and a profile of vertical movements measured across this cross section (lower inset). Cretaceous sedimentary rock has been thrust over Quaternary alluvial deposits on a fault dipping south. Benchmarks for releveling, among other instruments, were mounted in the tunnel. Repeated leveling during the late 1970s and early 1980s revealed the pattern of relative vertical movements shown in the inset in the lower right. The combination of vertical and horizontal measurements led to the logical inference that the scarp at the surface marked a thrust fault dipping south. A hole, drilled a depth of nearly 200 m to test this idea, confirmed the presence and dip of this fault. Diagonal lines parallel to the fault in Cretaceous rock indicate that the shear zone is some tens of meters wide. (Redrawn from Guseva [1986, Fig. 54, p. 102].)

landslide) on the north side of the range, as suggested by C. R. Allen (personal communication, 1990) and R. E. Wallace (personal communication, 1976), or some other superficial deformation, as suggested by Belokopytov and others (1992). One test of this alternative is to measure accurately the distances between points within the Peter the First Range with respect to points in its surroundings to the north and south. Toward this end, benchmarks installed on peaks and high points in and around the Peter the First Range were triangulated in 1966–1967 and again in 1974; then direct measurements of distances using laser-ranging devices began in 1974.

The uncertainties in the published measurements appear to be too large to resolve this question. The intervisible benchmarks are commonly 10 to 20 km apart. Guseva (1986, p. 28) reported rms errors in angles between them of 3 to 5×10^{-6}. Assuming typical errors of 6×10^{-6} in differences in angles between the two surveys, these errors translate into errors in relative positions of 60 to 120 mm for benchmarks 10 to 20 km apart and of 300 mm for those 50 km apart. Corresponding errors in average speeds would be 8 to 16 mm/a and 40 mm/a over the seven to eight year interval. Thus, rates of 10 mm/a should not be resolvable.

To define a reference frame for deformation, Guseva (1986, p. 64) used two benchmarks whose distance apart was measured using the laser-ranging device in 1966 and the azimuth between them. They lie at the west end of the network, north of the Peter the First Range. She then calculated relative speeds of other benchmarks with respect to this frame. Her calculated velocities show southeastward movement of the Peter the First Range increasing eastward from 20 mm/a to 60 mm/a over a distance of about 50 km. This increase could be an artifact of an error in an angle measured with respect to the reference line. At any rate, the velocities do not appear to be significant at the 95% confidence limit. It is worth noting that Guseva (1986, p. 150) reported uncertainties in velocities of a few to more than 30 mm/a, which are somewhat lower than those that we assume. Nevertheless, her reported velocities would not be significant at 95% confidence.

Guseva (1986, pp. 92–93) also analyzed these data in terms of local strains over the region, using groups of three benchmarks and the changes in the calculated distances between the three pairs of benchmarks for each group. The most rapid shortening occurred across the Vakhsh Overthrust, but the pattern of strain over the rest of the region is by no means homogenous or simple. These results offer no support for the idea that the Peter the First Range moves as an essentially rigid block with respect to the South Tien Shan, but because few of the calculated strains exceed 1×10^{-6}, very few of them would be resolvable given the rms errors in angles of 3 to 5×10^{-6}. Thus, these measurements do not seem to resolve the question of how much the Peter the First Range deforms.

In principle, the direct measurements of distances using laser-ranging devices should allow more accurate determination of the displacement or strain fields than do measurements using only angles, and a program to measure such distances began in 1974. Lengths of lines range from 7 to 22 km (with one at 32 km). Belokopytov and others (1992) reported uncertainties of 1×10^{-6}. Hence, uncertainties in distances should be about 10 to 30 mm, and uncertainties in average speeds between benchmarks over a period of six years should be only a few mm/a. Both Guseva (1986, p. 149) and, with additional data, Belokopytov and others (1992) reported rates of change in lengths from –15 mm/a to +38 mm/a. These relative movements, however, reveal no obviously consistent pattern, except that many of the most rapid rates are for the longest lines or for the shortest intervals of time between resurveys. As with the results of retriangulation, Guseva (1986, pp. 94–95) calculated strain rate tensors for groups of three benchmarks using changes in lengths between each pair. Large variations in both orientations and magnitudes of strain-rate over the region make interpreting the strain difficult. These results also offer no support for the idea that the relative movement of the Pamir, or the Peter the First Range, with respect to the South Tien Shan is localized at the Vakhsh Overthrust. The complexity of the apparent deformation, however, casts doubt on its accurately reflecting tectonic processes.

Summary of geodetic results. The geodetic observations pertain only to a small part of the Pamir region. Their importance results from the network spanning one of the major tectonic features, the Vakhsh Overthrust.

Taken as a whole the geodetic data do not permit an unambiguous interpretation. If one accepts Konopaltsev's (1971) results from the Runou Valley showing rapid convergence of the Peter the First Range with respect to the South Tien Shan and Guseva's (1986) data from the Sari-Pul' region showing rapid overthrusting at the Vakhsh Overthrust, one would logically conclude that the Peter the First Range is advancing northward onto the South Tien Shan at about 15 mm/a along a fault dipping moderately south, as corroborated by drilling. The releveling profiles (Fig. 33) suggest that this fault flattens with depth and to the south. On the other hand, these data do not rule out the possibility that much of the recorded strain results from surface processes (e.g., one or more large listric slumps) that are not deep-seated in origin. Moreover, the regional triangulation and trilateration over the region might also be interpreted in terms of chaotic relative movements of detached blocks of the earth's surface or of radially diverging upper crust of the Peter the First Range (Belokopytov and others, 1992; Lukk and Shevchenko, 1986).

Guseva and others (1983) argued that an interpretation in terms of surface processes is unlikely. First, other thrust faults are present in the region. Second, no major normal faults have been recognized in the range, which has been thoroughly studied geologically. Third, fault plane solutions of earthquakes consistently show northwest-southeast shortening of the region. These observations provide no support for the idea that the geodetic data reflect surface, instead of deep-seated, processes. Nevertheless, the data do not show unequivocally that the geodetically measured convergence at the Vakhsh Overthrust is an accurate measurement of the convergence of the entire Peter the First Range with respect to the South Tien Shan. We proceed assuming that the Peter the First Range does overthrust the South Tien Shan in a north-northwesterly direction at about 15 mm/a, but we recognize that this inference is not a unique interpretation of the entire geodetic data-set.

Partitioning of India's convergence with Eurasia across the Pamir

At the longitude of the Pamir, India moves nearly due north with respect to the stable parts of Eurasia at about 44 (±5) mm/a (DeMets and others, 1990). This convergence is not absorbed by underthrusting on one fault but is distributed from the southern edge of the Himalaya, across the Pamir, to the Tien Shan. The results summarized above are neither sufficiently complete nor sufficiently accurate to quantify the low rates of convergence on the minor faults, but they clearly suggest that roughly half of the convergence might be absorbed at the Trans-Alai Range along the northern edge of the Pamir.

A relatively accurate lower bound on the rate of convergence at the Himalaya is provided by a comprehensive study of the amount and timing of underthrusting of the Indian Shield beneath the Salt Range. Seismic reflection profiles allow the exposed geologic units to be extrapolated to depth, and these data indicate an offset on the thrust fault of 20 to 23 km (Baker and others, 1988; Lillie and others, 1987). Accurate dating, using magnetostratigraphy, of the initiation of deformation at the Salt Range implies that slip began between 2.1 and 1.6 Ma (Johnson and others, 1986). Thus, the average Quaternary rate of underthrusting is between about 9 and 14 mm/a (Baker and others, 1988). This rate is a lower bound for the whole of the Himalaya, because shortening could be occurring farther north within the Himalaya.

The Pamir seem to be overthrusting roughly north-south onto the Alai Valley and the South Tien Shan at a rapid rate of more than 20 mm/a, given the combination of rapid slip (10 to 15 mm/a) on the northeast– to north-northeast–trending Darvaz fault and the apparent rapid north-northwesterly convergence across the Garm region (roughly 15 ± 5 mm/a). This 20-mm/a rate of northward overthrusting, if correct and combined with that across the Salt Range, leaves only about 10 to 15 mm/a from the total convergence rate of 44 ± 5 mm/a between India and Eurasia (DeMets and others, 1990). Fault plane solutions of earthquakes show that some underthrusting of the Ferghana Basin southward beneath the South Tien Shan must occur (e.g., Nelson and others, 1987). Moreover, some shortening surely occurs on the northern margin of the Ferghana Basin. Although deformation within the Pamir does not appear to be rapid, there should be a few millimeters per year of shortening, given continued slip on the Aksu-Murgab fault system (no. 19 in Fig. 25). The Himalaya, too, may be undergoing internal crustal shortening at a few millimeters per year. Thus, conversely, either the convergence rate between India and Eurasia is greater than 44 ± 5 mm/a, or the rate of underthrusting of the South Tien Shan beneath the Pamir cannot be much more than 20 mm/a. Moreover, given that the rate for the Trans-Alai region is based on only a few crudely dated Holocene or late Quaternary offsets and on geodetic results that might be interpreted differently, perhaps we should allow for a rate of underthrusting lower than 20 mm/a. Nevertheless, relatively rapid localized convergence in the Trans-Alai region, surely at more than 10 mm/a and probably approaching 20 mm/a, seems inescapable.

INTERMEDIATE-DEPTH SEISMICITY, UPPER MANTLE STRUCTURE, AND EVIDENCE OF SUBDUCTED LITHOSPHERE

A zone of intermediate-depth earthquakes dips south-southeast beneath the Pamir (Figs. 23 and 35). The occurrence of intermediate-depth earthquakes is generally taken as evidence for the subduction of lithosphere, or of the sinking of cold material, into the asthenosphere. Nearly all intermediate- and deep-focus earthquakes occur in thin zones (thickness <30 km [Isacks and Barazangi, 1977]) where one slab of oceanic lithosphere clearly has been subducted in the last 10 to 20 m.y. Nevertheless, intermediate-depth earthquakes also occur in a few intracontinental settings where the evidence for recent subduction is equivocal, if not absent (e.g., Chen and Molnar, 1983). The most active intracontinental intermediate-depth seismicity occurs beneath the Pamir and the Hindu Kush. Because the earthquake hypocenters define relatively thin tabular zones, no wider than about 30 km, to depths of 150 to 300 km, they seem to mark slablike bodies of presumably cold material (e.g., Billington and others, 1977; Chatelain and others, 1980; Roecker and others, 1980) rather than the diffuse straining of a broad zone. Accordingly, like Hamburger and others (1992), we assume that these intermediate-depth earthquakes occur in slabs of lithosphere underthrust beneath the Pamir and the Hindu Kush.

Often the seismicity of the Pamir and Hindu Kush is treated as one continuous zone, a view that we doubt strongly. Nevertheless, we discuss both the seismicity and the deep structure of the neighboring Hindu Kush before exploiting only that of the Pamir.

The most intense intermediate-depth seismic activity in Central Asia occurs along a east-west zone beneath the northern edge of the Hindu Kush (Billington and others, 1977; Chatelain and others, 1980; Gutenberg and Richter, 1954; Lukk and Nersesov, 1970; Malamud, 1973; Nowroozi, 1971; Roecker and others, 1980). This zone dips steeply northward to a depth of more than 300 km and projects to the earth's surface just south of the Hindu Kush in Afghanistan. Like others, we presume that it results from northward subduction of lithosphere, but we cannot pinpoint the surface outcrop of the major thrust fault along which subduction occurs. It is note-

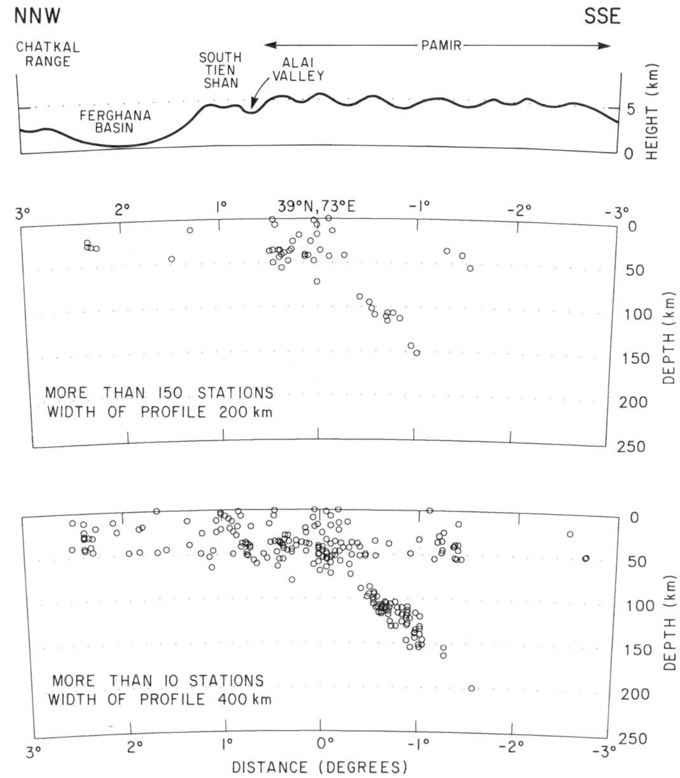

Figure 35. Cross sections of seismicity and topography through the Pamir showing a zone of intermediate-depth earthquakes dipping southeastward. Both profiles are oriented 145°, with a center at 39°N 73°E, and extending 3° to the northwest and southeast (Fig. 2). Earthquakes with epicenters within 100 km of this line are plotted in the upper profile and within 200 km in the lower one. The bottom profile shows all earthquakes located by the International Seismological Center between 1964 and 1985 with more than 10 stations reporting P-wave arrival times, and the upper cross section shows only those hypocenters located with P-wave arrival times from 150 stations or more. Notice that the inclined zone dipping south-southeast is well defined and reaches a depth of 150 km. With the less reliable locations, it seems to reach 200 km. The projection of this zone intersects the surface at the northern margin of the Trans-Alai Range and requires a downdip length of at least 250 km and probably of 300 km or more.

worthy that whereas the deformation between India and Eurasia is spread over a broad zone along most of the Himalaya and farther north, the north-south dimension of the zone of high mountains and Cenozoic deformation is particularly small west of the Pamir (Figs. 2 and 24), where intermediate-depth seismicity is most intense (Fig. 23). This obviously is consistent with the particularly narrow width of the Hindu Kush being a consequence of localized subduction of lithosphere instead of the distributed crustal shortening that created the much wider Pamir and Tibetan Plateau farther east.

With the inclusion of relatively poorly located earthquakes, the Hindu Kush zone seems to continue to the northeast beneath the western and central parts of the Pamir (Fig. 23) Precisely located earthquakes, however—those recorded well by many local stations—reveal a gap between the seismic zones beneath the Hindu Kush and the Pamir (Chatelain and others, 1980; Roecker and others, 1980). Seismicity beneath the Pamir defines a zone that dips south-southeastward (Fig. 35) (Billington and others, 1977; Lukk and Nersesov, 1970; Malamud, 1973). Earthquakes are less frequent in the Pamir than in the Hindu Kush zone, but they are sufficiently numerous that the width of less than 30 km, the dip of about 45°, and the downdip length of nearly 300 km are well constrained (Fig. 35).

One simple argument suggests that the seismic zone beneath the Pamir is similar to those occurring at island arcs. The length of such a seismic zone is related to the rate of subduction and the (thermal) age of the subducted lithosphere by

$$\text{length of seismic zone} = (\text{rate of subduction} \times \text{age of lithosphere}) / 10$$

(Molnar and others, 1979). For a subduction rate of about 20 mm/a and a Jurassic age (about 150 Ma) of thermally perturbed lithosphere (Leith, 1985), the calculated length is 300 km.

The Pamir and Hindu Kush zones are also associated with very low attenuation of seismic waves. This is clear not only from measurements of the frequency content of seismic waves from earthquakes to nearby stations (Khalturin and others, 1977; Roecker and others, 1982) but also from intensity distributions of the larger events (Lukk and Nersesov, 1970). Inclined seismic zones at island arcs are commonly associated with low attenuation and are presumed to define cold slabs of lithosphere immersed in hot asthenosphere where attenuation is high (e.g., Oliver and Isacks, 1967).

Travel times from distant earthquakes recorded by stations near the Pamir–Hindu Kush region reflect systematic variations that imply marked lateral variations in velocity in the mantle of this region (Vinnik and Lukk, 1973, 1974; Vinnik and others, 1977, 1978). Rays passing through the seismic zone arrive earlier than those coming from other directions, suggesting high seismic wave velocities in the region of the intermediate-depth seismic zone. Such high velocity zones are well established by rays passing through inclined seismic zones at island arcs (e.g., Mitronovas and Isacks, 1971). Accordingly, Vinnik and others (1977) suggested that thick lithosphere from a continental shield had been underthrust in is area.

A detailed investigation of the seismic wave velocities in the upper 200 km of the mantle near the Hindu Kush, however, revealed a surprise: lower P- and S-wave velocities near the seismic zone than outside it (Roecker, 1982). As mentioned above, Roecker inferred that not only mantle lithosphere but also some continental crust had been subducted to at least 150 km beneath the Hindu Kush. His data are inadequate to resolve lateral variations in velocity at greater depth or farther north beneath the Pamir. Thus, even if Roecker's

(1982) inference about subduction of continental crust were correct for the Hindu Kush, his data do not apply to the Pamir. Moreover, they permit the possibility that such subducted continental crust followed oceanic lithosphere and that Vinnik and Lukk's (1973, 1974) inference of high velocity applies to greater depths. Recent, preliminary work of Mellors and others [1991] suggests compatibility of Roecker's and Vinnik and Lukk's interpretations for their respective depths.

In summary, planar zones of intermediate-depth earthquakes dip at about 45° southward beneath the Pamir to a depth of 150 to 200 km (Fig. 35) and steeply northward beneath the Hindu Kush to a depth of more than 300 km. These seismic zones seem to be associated with zones of low seismic wave attenuation and, at least in their deeper parts, with high seismic wave velocity (Vinnik and Lukk, 1973, 1974) as is common at island arcs where intermediate and deep earthquakes occur in cold slabs of subducted oceanic lithosphere. Hence, these seismic zones seem to define two down-going slabs of lithosphere: south-southeastward beneath the Pamir and north-northwestward beneath the Hindu Kush (Fig. 36). The intermediate-depth earthquakes in the Hindu Kush region, however, seem to occur very near a low velocity region, which could be subducted continental crust (Roecker, 1982). The Pamir zone is the much less thoroughly studied, and the presence of cold material is revealed only by the intermediate-depth seismicity. The tabular configuration of this seismic zone, however, suggests that a slab of lithosphere has been subducted. If that subduction occurred at the outer margin of the Pamir, the length of the slab must be at least 250 to 300 km (Fig. 35). Its projection to the surface suggests that convergence is localized along the northern margin of the Pamir (Hamburger and others, 1992), where active faulting seems to be more intense.

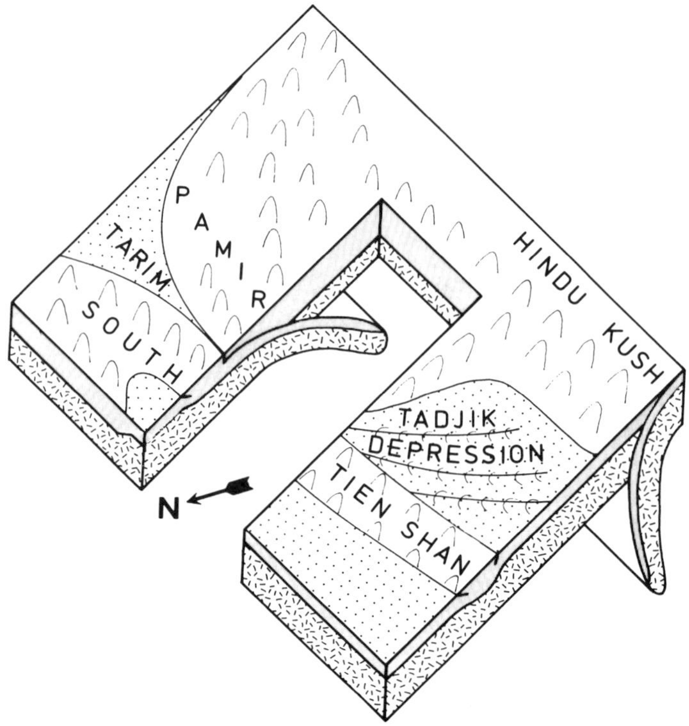

Figure 36. Block diagram illustrating lithospheric structure of the Pamir–Hindu Kush region. A slab of lithosphere has been underthrust south-southeastward beneath the Pamir. A different slab has been underthrust north-northwestward beneath the Hindu Kush. Crust is shaded, and mantle lithosphere is denoted by irregular hatching. Dots show areas with thick Cenozoic sediment surrounding mountainous terrain.

SUMMARY AND SYNTHESIS

Our objective is to determine how much convergence between the southern Pamir and the stable parts of Eurasia has been absorbed by underthrusting of continental lithosphere into the asthenosphere. To make such a determination requires that we make quantitative estimates of the amounts of convergence absorbed by other processes.

Three independent observations (see "Geological Constraints on the Magnitude of Convergence Between the Pamir and the rest of Eurasia") imply that in Cenozoic time the northern Pamir has advanced onto an area that lay immediately to its north. First, the late Paleozoic suture in the Northern Pamir appears to be a continuation of the same suture that crosses the Hindu Kush in northern Afghanistan and that follows the Kunlun in northern Tibet (Figs. 3 and 4); hence it has been displaced roughly 300 km northward from its eastern and western continuations. Second, paleomagnetic declinations in Cretaceous and Paleogene sedimentary rock along the outer margin of the Pamir show large anomalies that imply large rotations about vertical axes (Fig. 5); these rotations can be interpreted as resulting from a northward displacement of 300 to 700 km of the outer margin of the Pamir with respect to its continuation through the Kopet Dagh farther west (Fig. 6). Third, east-west–trending facies zones within Cretaceous and Paleogene sedimentary rock in the Tadjik Depression are abruptly truncated at the Vakhsh Overthrust, and their eastward continuations have been displaced a minimum of about 200 km northward (Figs. 14, 15, and 16) (Suvorov, 1968). With as much as 115 km of shortening within the outer zone of the Pamir, convergence of 300 km (or more) between the Northern Pamir and Eurasia is implied. Thus, collectively, these observations imply that the Northern Pamir has been displaced northward about 300 km or more with respect to the Asian crust farther north. Structural cross sections across the Vakhsh and Trans-Alai overthrusts (Figs. 17, 19, and 20) show them to be major features but place no useful lower bound on the amount of shortening there.

Although the rocks of the Pamir have been assembled by collisions in late Paleozoic and in Mesozoic time, deformation of Cretaceous and Tertiary rock attests to Cenozoic crustal shortening. In some cases klippen allow estimates of large displacements of thin thrust sheets, and in others offsets on strike-slip faults that terminate at thrust faults refine estimates of thrust displacements (Figs. 8, 9, 11, 12, and 13). A minimum of 300 km of shortening *within* the Pamir implies that the southern edge of the Pamir has moved northward 600 km or more with respect to the area farther north.

Some of this convergence between the Karakorum and the South Tien Shan must have been absorbed by crustal thickening within the Pamir. Scattered outcrops of Cretaceous marine carbonate sediment in the Central Pamir (Shvol'man, 1977) attest to low elevations at that time, and it seems logical to assume that the present high elevations were reached in Cenozoic time following India's collision with Eurasia. Thus, the present thickness of about 70 (± 10) km implies a doubling of the crust in Cenozoic time. Given a present north-south width of about 300 km for the Pamir and an assumed initial thickness of 35 (± 5) km, this implies that of the 650 km or more of convergence between the Southern Pamir and the South Tien Shan, about 300 (± 100) km has been accommodated by crustal shortening and thickening.

We infer that the crust that occupied the area presently encompassed by the Pamir was thinner than 35 (± 5) km before the Pamir overrode this region. The Tadjik Depression is covered by a thick sequence of Jurassic to Paleogene sedimentary rock that in turn is overlain by a thick sequence of Neogene terrigenous deposits. The mean thickness of sedimentary rock exceeds 10 km, and the thickest deposits are in the southeastern part of the depression where the Moho is deepest (e.g., Bekker and others, 1974a; Kulagina and others, 1974). The thickness of crystalline basement of the Tadjik Depression appears to be only 20 to 25 km. The relatively thick Jurassic and Paleogene deposits accumulated in a basin whose subsidence followed an exponential decay, which Leith (1985) interpreted as evidence for cooling and contraction of a thinned lithosphere. If this sedimentary basin continued eastward to the Tarim basin, then relatively thin crust characterized the area now occupied by the Pamir, prior to India's collision with Eurasia.

Although the Tadjik Depression is by no means inactive, and as much as 100 km of northwest-southeast shortening may have occurred across it in Cenozoic time, virtually all of this deformation appears to be confined to the sedimentary cover, above a Tithonian layer containing salt and gypsum. Moreover, the presence along the outer margin of the Pamir of Cretaceous and Paleogene sedimentary rock that was deposited 200 to 300 km farther south along the southern margin of the Tadjik sedimentary basin (Figs. 14, 15, and 16) suggests that the overriding of the Pamir onto that basin scraped off this cover and pushed it northward. The thin crystalline basement on which that sediment was deposited must underlie the Pamir, but it cannot contribute much to the thick crust there.

The active tectonics, as revealed by seismicity, Quaternary faulting, and geodetic estimates of present-day crustal movements, imply that roughly half of India's 44 ± 5 mm/a northward penetration into Eurasia (DeMets and others, 1990) is absorbed near the outer margin of the Pamir, along the Trans-Alai Range and Vakhsh Overthrust. Seismicity is higher along this zone than in the rest of the Pamir or Tadjik Depression (Hamburger and others, 1992) (Figs. 23 and 24). More importantly, the combination of 10 to 15 mm/a of left-lateral slip along the north-northeast–trending Darvaz fault (Fig. 25) and the roughly 10 to 15 mm/a of geodetically determined, north-south shortening across the Peter the First Range northwest of the Darvaz fault (Figs. 32 and 33) suggests roughly 20 mm/a of north-south shortening along the Pamir-Alai zone, east of where the Darvaz fault and Peter the First Range meet. Thus, active convergence seems to be absorbed by localized underthrusting along this belt (Hamburger and others, 1992).

The zone of intermediate-depth earthquakes, which dips southeastward beneath the Pamir at about 45° (Fig. 35), also projects to the earth's surface near the Pamir-Alai zone (Hamburger and others, 1992). Because intermediate-depth seismicity seems to occur in cold slabs of lithosphere that have plunged into the hotter asthenosphere in the last 10 to 20 million years, it is logical to assume that the lithosphere in which these intermediate-depth earthquakes occur was underthrust along the Pamir-Alai zone.

Thus, the budget of convergence across the Pamir implies that the intermediate-depth earthquakes occur in lithosphere that once underlay the area now occupied by the Pamir (Figs. 37 and 38). Recall that the northern margin of the Pamir seems to have overridden more than 300 km of southern Eurasia. Crustal shortening within the Pamir has also allowed another more than 300 km of convergence of the southern edge of the Pamir with respect to Eurasia. Thickening of crust initially of normal thickness can account for only 300 ± 100 km of that shortening and convergence (Figs. 37 and 38). The length of the intermediate-depth seismic zone also implies that about 300 km of cold lithosphere has been subducted. That subducted lithosphere must have underlain the sedimentary rock that is now folded in the Pamir-Alai zone and that was scraped off the crust that once lay east of the Tadjik Depression. By analogy with the crust beneath the Tadjik Depression, we infer that the crustal thickness was 20 to 25 km.

Probably the weakest link in the logic presented above is the part that will be most difficult to test: the inference of the thickness, and nature, of crust that was subducted (e.g., Tapponnier and others, 1981). None of that crust remains near the surface. We cannot rule out the possibility that a small ocean basin, like that beneath the Black Sea or possibly beneath the Caspian Sea, occupied the territory where the Pamir now lies (Chatelain and others, 1980; Legler and Przhiyalgovskaya, 1979; Leith, 1985; Tapponnier and others, 1981). If so, the intermediate-depth earthquakes would occur in the oceanic lithosphere of that basin. Such an explanation, however, requires a fortuitous decrease in crustal thickness from 20 to 25 km beneath the Tadjik Depression to a much smaller value where no such crust remains to be studied. Moreover, if the basin deep-

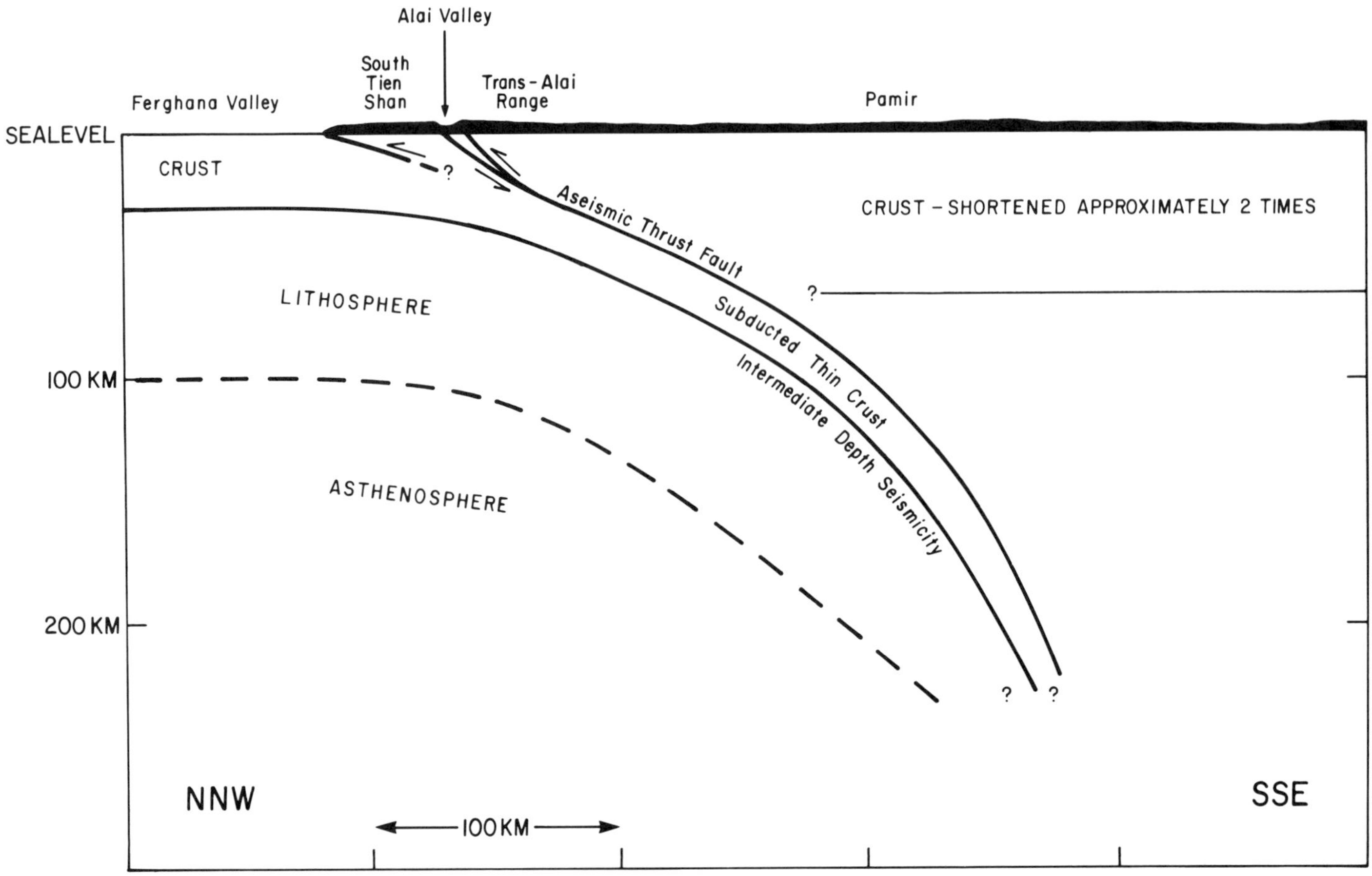

Figure 37. Simple interpretive summary cross section across the Pamir. Asian lithosphere, of roughly normal thickness, is shown underthrust beneath the Pamir to a depth of at least 200 km. Crustal thickening within the Pamir, by thrust faulting at relatively shallow depths but at greater depths by processes unconstrained by existing data, has produced a thickness roughly twice normal. Some underthrusting of the Ferghana Basin must occur beneath the South Tien Shan, where upper crust is probably being sheared over the lower crust. We suspect, however, that most of the crust subducted to depths greater than 100 km was thinner than that exposed at the surface along this profile.

ened eastward from the Tadjik Depression, facies zones of Cretaceous and Paleogene deposits would trend north-south, not east-west. These arguments do not categorically eliminate the possibility that oceanic lithosphere was subducted beneath the Pamir, but subduction of 300 km of relatively thin continental crust (20 to 25 km) beneath the Pamir is clearly the simplest interpretation.

APPENDIX A. DISCUSSION OF EARTHQUAKES IN THE PAMIR REGION

The following appendix summarizes the analysis of the 13 earthquakes in the Pamir region that we have studied. Using McCaffrey and Abers's (1988) program, which implements methods developed by Langston and Helmberger (1975) and Nábělek (1984), we calculated synthetic waveforms from point sources embedded in a half-space with P- and S-wave velocities of 6.2 and 3.6 km/s and a density of 2,800 kg/m^3. The fault plane solution of the point dislocation is described by the strike, dip, and rake of one of the two nodal planes (Table 3). A series of overlapping pulses of triangular shape and half-widths of 1 second parameterize the time history of slip. McCaffrey and Abers's (1988) program solves for the three angles describing the fault plane solution, the focal depth, the relative weights of the source time function elements, and the seismic moment by minimizing, in a least square sense, the differences between synthetic and recorded waveforms. We used this procedure with the waveforms weighted inversely with the azimuthal density of stations

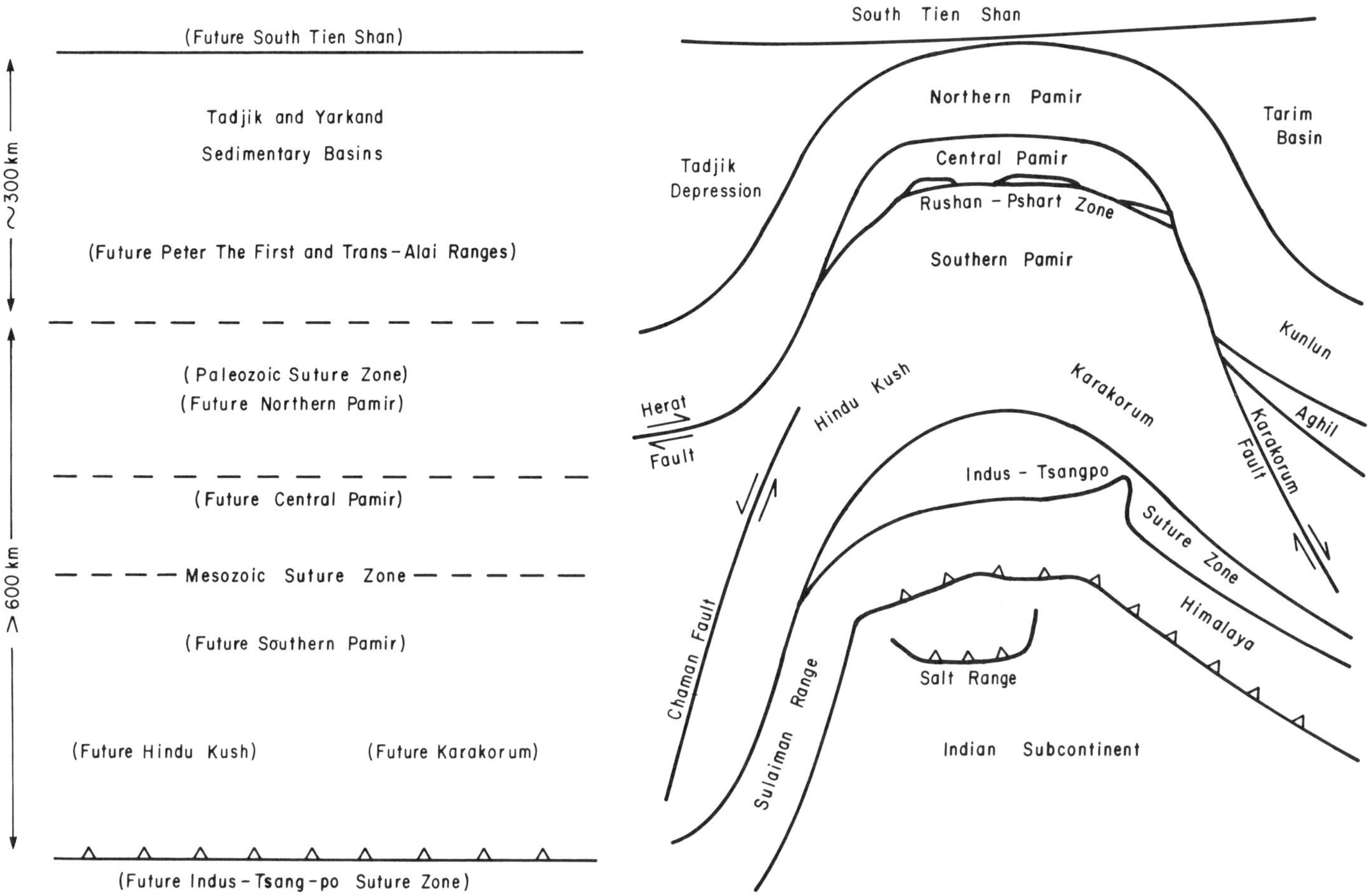

Figure 38. Simplified maps showing budgets of crust involved in the formation of the Pamir. The left side shows the approximate configuration before India's collision with Eurasia. The Northern Pamir seems to have been displaced northward 300 km (or more) over the area between the Tadjik and Yarkand basins. The Mesozoic suture zone, marked now by the Rushan-Pshart zone, has advanced northward with respect to the northern edge of the Northern Pamir by crustal shortening within the Northern and Central Pamir, and the Indus-Tsangpo suture zone has advanced north with respect to the Rushan-Pshart zone by shortening within the Southern Pamir. Such shortening exceeds 300 km. Before the collision, the northern edge of the Indian Plate lay at the Indus suture zone. Now, after slices of India's crust have been thrust atop one another to make the Himalaya, the surface outcrop of this boundary lies south of the suture and along the southern edge of the Himalaya, or within the Salt Range.

and with P waveforms weighted twice those of SH. Then, beginning with the minimum misfit, we searched for source parameters that yielded slightly worse, but nonetheless subjectively acceptable, misfits. (See Molnar and Lyon-Caen [1989] for further discussion about the approach taken here.)

For most of these events we show two figures. The following discussion summarizes what they show.

For nearly all events, one figure, generally labeled a, shows the comparisons of recorded (solid lines) and synthetic (dashed lines) waveforms for P and SH surrounding their respective focal spheres. Above the focal sphere, the date and for 1974 August 11 the hour define the earthquake (Table 3), and the basic source parameters are listed as strike/dip/rake/depth/seismic moment, where the first three are in degrees, the fourth in kilometers, and the fifth in units of 10^{16} Nm. For earthquakes whose P-wave first motions have not been plotted elsewhere, we show compressional first motions as closed symbols and dilatations as open symbols. P- and T-axes are shown by a large open circle with an X in the center and by a large closed circle, respectively. A time scale for the seismograms and the source time function are plotted near the P-wave focal sphere. Amplitude scales for P and for SH, normalized to a distance of 40° and a magnification of 3,000, are shown for each. Stations are labeled by 3- or 4-letter

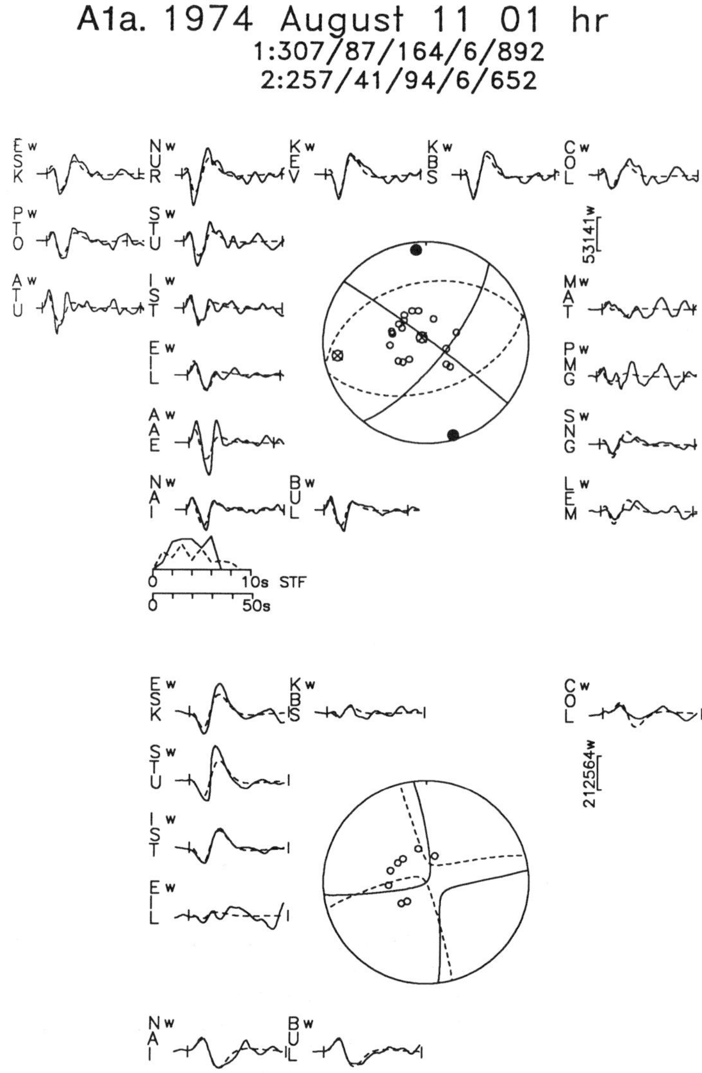

codes, and the lower case letters *w* or *d* indicate analogue seismograms of the WWSSN or digital seismograms of the Global Digital Network.

For most earthquakes a second figure, usually b, shows comparisons of synthetic and recorded waveforms for six stations and for a variety of fault plane solutions for which the fits at some stations are not very good. For each of these, the top row shows comparisons for the "best"-fitting solution. The next three pairs of rows show poorer fits, for limiting values of the strike, the dip, and the rake. They are meant to provide readers with information that gives them an impression of the uncertainties in these three angles.

For the three largest earthquakes, those of 1974 August 11 (01 hr), 1978 November 1, and 1985 August 23, obtaining acceptable fits of synthetic and recorded waveforms required searching among multiple events with relatively long source time functions. Consequently, establishing uncertainties by the same procedure was difficult, and we present diagrams designed to demonstrate the existence of multiple sources.

Using P-wave first motions, Jackson and others (1979) determined the fault plane solution for the event of 1967 May 11, 39.41°N, 73.77°E, depth = 11 ± 3 km, strike = 132° ± 10°, dip = 90° ± 20°, rake = 169° ± 15°. This is a small earthquake, and the P phase at JER is the only signal large enough to digitize. The match of the synthetic seismograph to it implies a focal depth of 11 ± 3 km. The P-wave first motions show largely right-lateral strike slip faulting on a northwesterly trending plane or conjugate faulting. We simply assumed such faulting and let the match of the P phase at JER refine the value of the rake.

1974 August 11 (01:13) Markansu Valley earthquake (Fig. A1). 39.38°N, 73.81°E. If one source depth = 2 to 3 km, strike = 291°, dip = 62°, rake = 131°. If two sources, depths < 10 km, strikes = 307° and 257°, dips = 87° and 41°, rakes = 164° and 94°.

This earthquake is puzzling, and its source parameters are poorly constrained. P-wave first motions cannot be fit by any pair of orthogonal nodal planes (Balakina, 1983; Jackson and others, 1979). Ni (1978) inferred oblique thrust faulting, similar to the point source double couple that "best" fits the P and SH waveforms (Fig. A1b), as discussed below. Jackson and others (1979) suggested a large strike-slip component (right-lateral on a northwest-trending plane) from the bulk of the P-wave first motions and from the S-wave first motions. Langston and Dermengian (1981), however, showed that the P waveforms fit such a solution poorly and that a thrust solution with a steeply northward dipping plane and a gently southward dipping plane matched them quite well. That solution, however, violated a number of strong P-wave first motions at stations southeast of the earthquake (Balakina, 1983; Molnar and Jackson, 1981). With some first motions violated by all solutions, the source of this earthquake clearly must be more complicated than can be represented by a single, point-source double couple. Yet, finding an acceptable solution has proven to be very frustrating, and eventually we used two sources (Fig. A1a).

We started by seeking the "best" point-source double couple that we could find to match those P and SH waveforms that we could digitize. McCaffrey and Abers's (1988) inversion routine consistently returned solutions that looked very much like Ni's (1978), consisting of planes striking east-northeast and east-southeast with reverse components and with right-lateral slip on the east-southeast–trending plane (see the second row of Fig. A1b). The east-plunging B-axis is required by the larger P-wave amplitudes at stations to the west than those to the east. SH waveforms are not very sensitive to the orientations of the planes, because they are very similar for pure strike-slip faulting like that of Jackson and others (1979), for Langston and Dermengian's (1981) pure thrust solution, and for Ni's (1978) oblique thrust solution. The relative amplitudes of P and SH waveforms rule out a pure strike-slip solution and favor the oblique-slip solution.

In searching for a good fit to the digitized seismograms, we found no way to match the small initial P-wave arrivals at stations in northern Europe and at COL. In fact many of these initial signals are

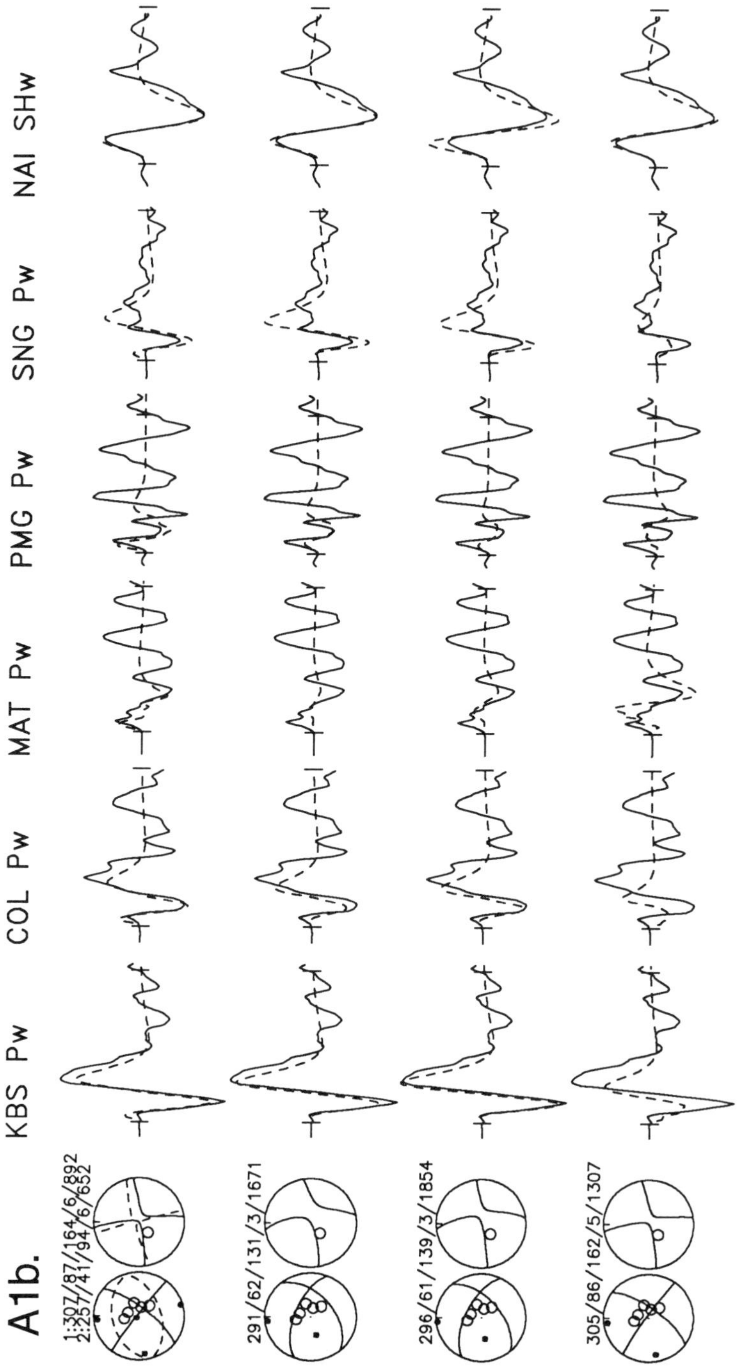
A1b.
KBS Pw
COL Pw
MAT Pw
PMG Pw
SNG Pw
NAI SHw
1:307/87/164/6/892
2:257/41/94/6/652
291/62/131/3/1671
296/61/139/3/1854
305/86/162/5/1307

too small to be seen on the digitized seismograms without enlarging them. Clearly if a small subevent produced them, it was much smaller than the main event that produced the larger signals the were fit by Langston and Dermengian (1981) or in Figure A1a. Apparently, Molnar (who among Jackson and others [1979] studied the P-wave first motions) paid more attention to these small precursors than they deserved. Nevertheless, there are other aspects of the solution with oblique-slip faulting that are not very satisfying.

First, the "best"-fitting focal depth was 2 to 3 km, very shallow for an earthquake with a seismic moment between 1 and 2×10^{19} Nm. Indeed, other earthquakes with comparable seismic moments have occurred in Australia at such shallow depths (Fredrich and others, 1988; McCaffrey, 1989), but those Australian earthquakes were associated with very clear surface ruptures. As discussed in the text, no clear surface faulting was found in the epicentral area of the Markansu Valley earthquake. The east-west dimension of the aftershock zone is about 20 km (Jackson and others, 1979). If the focal depth were only 2 to 3 km, so that the rupture were confined to the upper 5 km, the rupture area would have been only about 150 km^2, and a seismic moment of 1.5×10^{19} Nm would imply an average displacement of 3 m. It is difficult to imagine that such a displacement would not have resulted in a prominent surface break. A greater centroid depth of focus, however, yields a smaller average displacement and is more consistent with the absence of surface faulting.

Second, as shown in the second row of Figure A1b, no point-source double couple yielded a good fit to the P waveforms at MAT or PMG, which lie to the east of the epicenter. We searched among various fault plane solutions, including those in which weights assigned to these two seismograms were 10 times greater than those of the others. Different combinations of weights yielded solutions that differed by only about 10° in their strikes and rakes from the "best"-fitting point source. We can only conclude that a point source double couple is inadequate to describe the source of this earthquake. Such an inference is a sensible one given both the small precursors described above and the different fault plane solutions for each of the four largest shocks in the sequence (Jackson and others, 1979).

Before trying to match the waveforms with two solutions, we carried out another experiment. We inverted the P waveforms with the SH waveforms weighted 100 times less (third row in Fig. A1b), and we did the same with the SH waveforms weighted more by 100 times (fourth row in Fig. A1b). The P-wave solution included a slightly larger thrust component than the "best"-fitting solution, and the SH solution yielded a larger strike-slip solution. Both, of course, fit the phases with small weights worse than does the "best"-fitting solution, but neither matched the P waveforms at MAT and PMG well.

Thus, we proceeded assuming that a plausible solution should consist of two subevents, one largely strike-slip and the other nearly pure thrust. For such a combination, P-wave amplitudes toward the east or west would be reinforced, but those to the north would destructively interfere. SH phases would constructively interfere in all directions. With two subevents, however, the number of free parameters more than doubles, and the inversion routine became unstable. Thus, we arbitrarily fixed the locations and origin times for the two subevents to coincide, and we fixed the depths on individual runs. For each trial depth, the inversion routine returned strikes, dips, rakes, seismic moments, and source time functions for each subevent. The minimum least-squares misfit was found for a depth of 5 km, but the matches of synthetic and recorded waveforms for depths of 4 or 6 km were not noticeably different. The resulting pair of fault plane solutions yields acceptable fits of P waveforms at MAT and PMG (top row in Fig. A1b), as well as detectably better fits at SNG and LEM and slightly better fits at other stations than the "best"-fitting double couple (second row in Fig. A1b).

A2a. 1974 August 11 20 hr
267/55/81/6/44

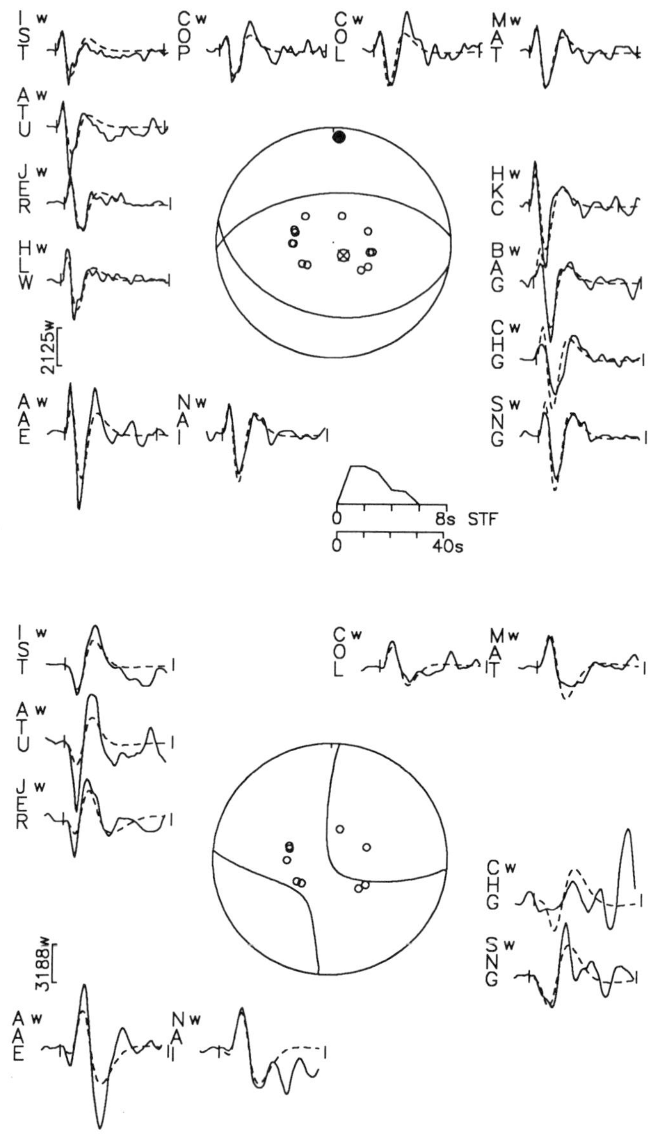

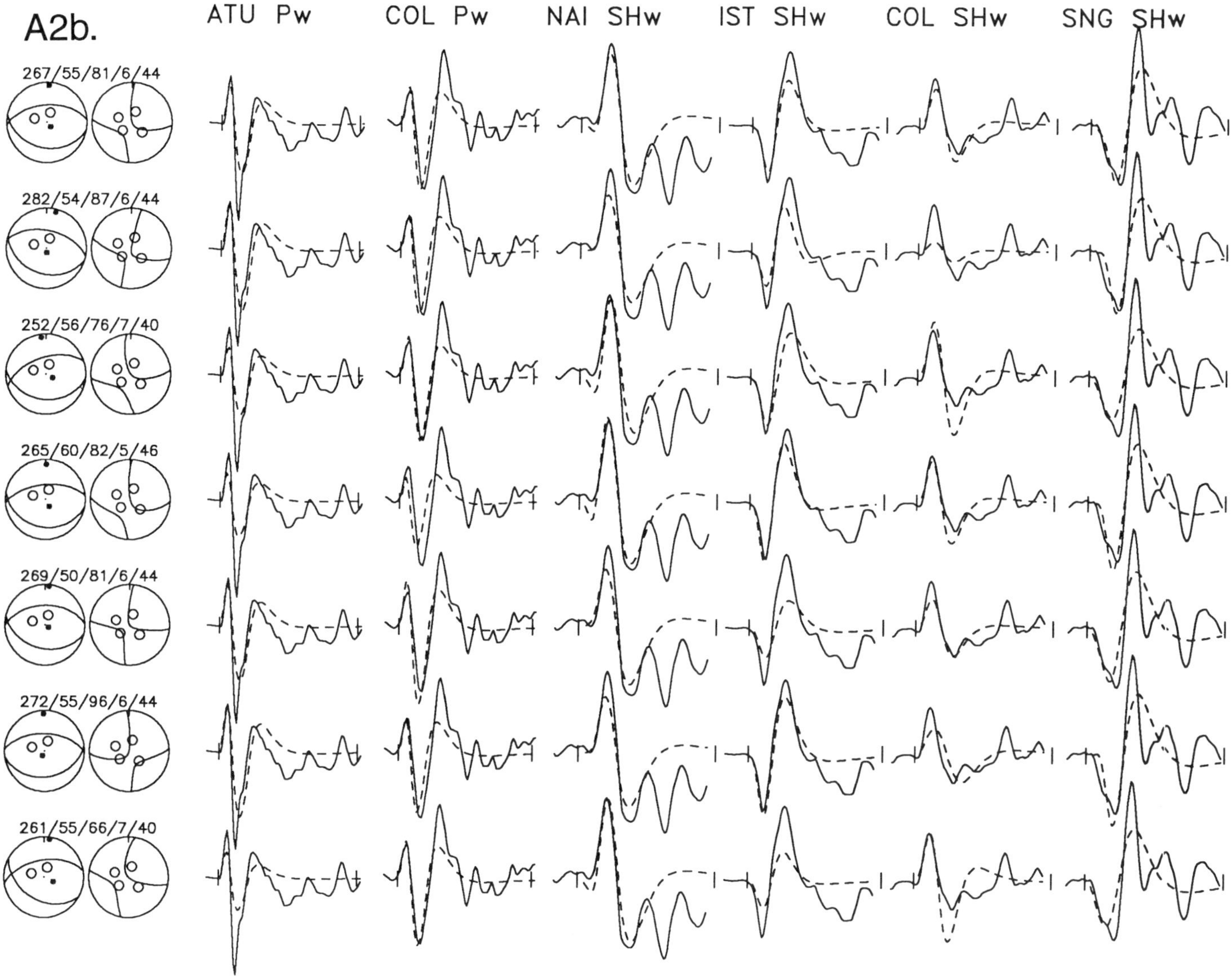

With so many free parameters, it seemed pointless to try to assign uncertainties to them. In fact, the improvement gained with two sources is significant at only two stations. Moreover, the contribution to the regional strain field by the two subevents is very similar to that of the "best"-fitting single event. Thus, we conclude that the Markansu Valley mainshock was associated with a large component of north-south crustal shortening, which was partitioned roughly equally into crustal thickening and east-west extension.

1974 August 11 (20:05) (Fig. A2). 39.47°N, 73.66°E, depth = 6 ± 3 km, strike = 267° ± 15°, dip = 56° ± 5°, rake = 81° ± 15°.

This was the first major aftershock of the Markansu Valley earthquake, which occurred earlier in the same day. The relocated epicenter lies 10 km north of that of the mainshock (Jackson and others, 1979). Both first motions and P and SH waveforms require a large thrust component (Fig. A2a), either on a steeply northward dipping plane or a gentler southward dipping plane (Balakina, 1983; Jackson and others, 1979). We use the steeper plane to describe the fault plane solution, because the description of uncertainties is more stable for it, but probably the southward-dipping plane was the fault plane. In any case, none of the suggested planes that ruptured in the mainshock could have been the fault plane for this event.

The waveforms, particularly at COL, allow uncertainties to be assigned to the nodal planes (Fig. A2b). For a strike of 282°, 15° clockwise from the "best"-fitting value of 267°, synthetic SH phases at JER, ATU, IST, and COL are too small. For 252°, the synthetic P phases at ATU and MAT are a little too small and a little too big, respectively, and synthetic SH phases are too small at HLW and ATU and too big at COL. For a dip of 60°, 5° steeper than the "best"-fitting value, synthetic P phases at ATU, HLW, JER, IST, COP, and COL are all too small, and the sSH at COL is too big. For a dip of 50°, synthetic P phases at COL and MAT are a little too big, and synthetic SH phases at NAI, AAE, JER, ATU, IST, COL, and MAT are too small. For a rake of 96°, 15° steeper than the "best"-fitting 81°, the synthetic P phase at NAI is a little too big, and the SH phases at COL and at CHG are too small and too big, respectively. For a rake of 66°, synthetic P phases at ATU, IST, and COP are too small; SH phases at JER, ATU, and IST are too small; and sSH is too big at COL.

1974 August 11 (21:21) (Fig. A3). 39.49°N, 73.65°E, depth = 10 ± 3 km. Strike = 62° + 20°/- 15°, dip = 75° + 10°/- 8°, rake = 323° ± 10°.

The seismograms and the fault plane solution differ markedly from those of the previous aftershock, which occurred only an hour earlier (Balakina, 1983; Jackson and others, 1979). The clear dilatations at stations to the north and west are suggestive of a significant component of normal faulting, but the P phases at BAG and HKC and the SH phases require a large strike-slip component (Fig. A3a). We describe the solution and its uncertainty in terms of the northeast-trending plane, but we suspect that right-lateral strike-slip displacement occurred on the north-northwesterly striking plane.

The double-peaked character of the seismograms at some stations suggests a multiple event might have occurred, but repeated efforts to match the seismograms with a multiple event failed. The inversion routine consistently returned a simple time function with a depth near 10 km. If the source was complex, then the subevents must have ruptured quite different planes.

The match of the synthetic seismograms to the recorded waveforms was disappointing in that no solution matched all of them well, and it is clear that if we had fewer waveforms, we might have deduced a different solution. This is illustrated by examples of solutions deemed poor by fits at several stations, but for which fits were better at other stations (Fig. A3b).

The strike of the east-west–trending nodal plane is not constrained well, corresponding to an uncertain rake on the other plane. Large synthetic P phases at MAT, SNG, KEV, and KBS and the small synthetic P at BAG and SH at COL render unacceptable a strike of 82°, 20° counterclockwise from the "best" solution, but the SH phases at SNG and AAE, fit poorly for the "best" solution, are fit well. In contrast, a strike of 47°, 15° clockwise from the "best" solution, makes the poorly fitting initial phases of SH at SNG and AAE worse as well as synthetic P phases at HLW, JER, IST, COP, GDH, COL, and SNG that are a little too big. A relatively steep dip of 85°, corresponding to nearly pure strike-slip displacement on the north-northwesterly–trending plane, yields synthetic P phases at HKC and BAG that lack the double pulses of the recorded waveforms. In addition, the synthetic P phase at SNG is too big, and the synthetic sSH at COL is also too big, but again the SH phases at SNG and AAE are fit somewhat better than for the "best" solution. For a relatively gentle dip of 67°, the P-wave first motions at HKC and BAG are opposite to those recorded, and synthetic P phases at HLW, JER, KEV, KBS, GDH, COL, and BAG fit poorly. For a relatively gently plunging rake of 338°, corresponding to a gentler dipping north-northwesterly–trending plane than the "best" solution, the initial motion of

A3a. 11 August 1974 21 hr
62/75/328/10/59

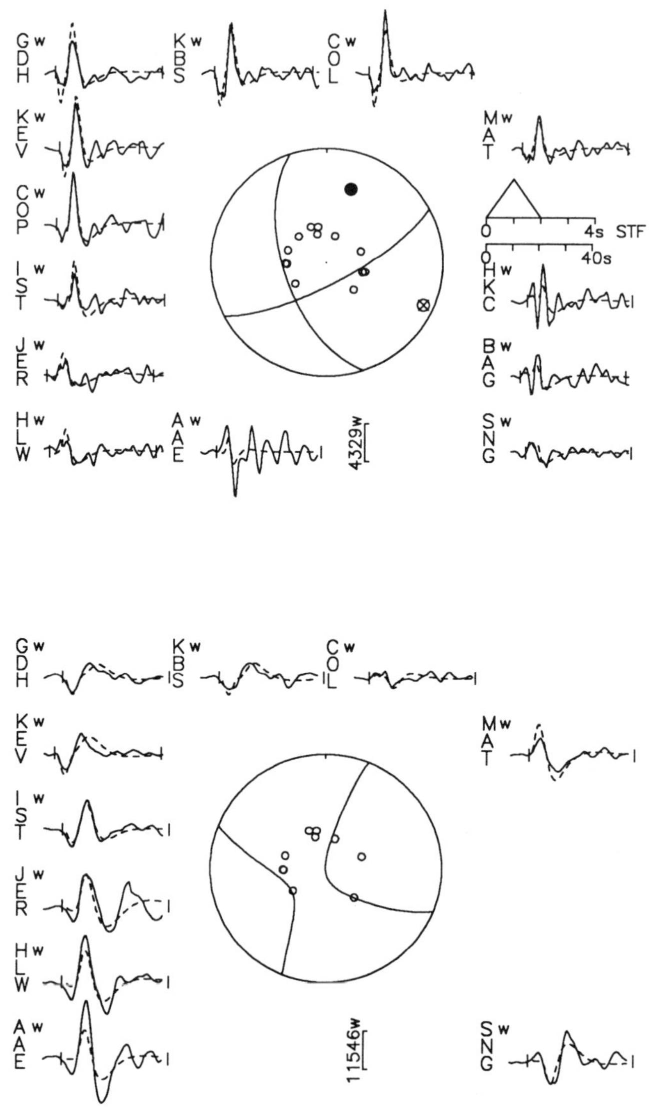

synthetic P phases at JER is opposite that recorded, the P at IST is too small, the synthetic SH phases at KEV and GDH are a little too big, and the initial phases of the synthetic SH phases at SNG and AAE are opposite to those observed, but the fits of P phases at HKC and BAG are the "best" that we obtained. For a steeper rake of 318°, corresponding to a gentler dip of the north-northwesterly–dipping plane than for the "best" solution, synthetic P phases at HLW, JER,

A3b.

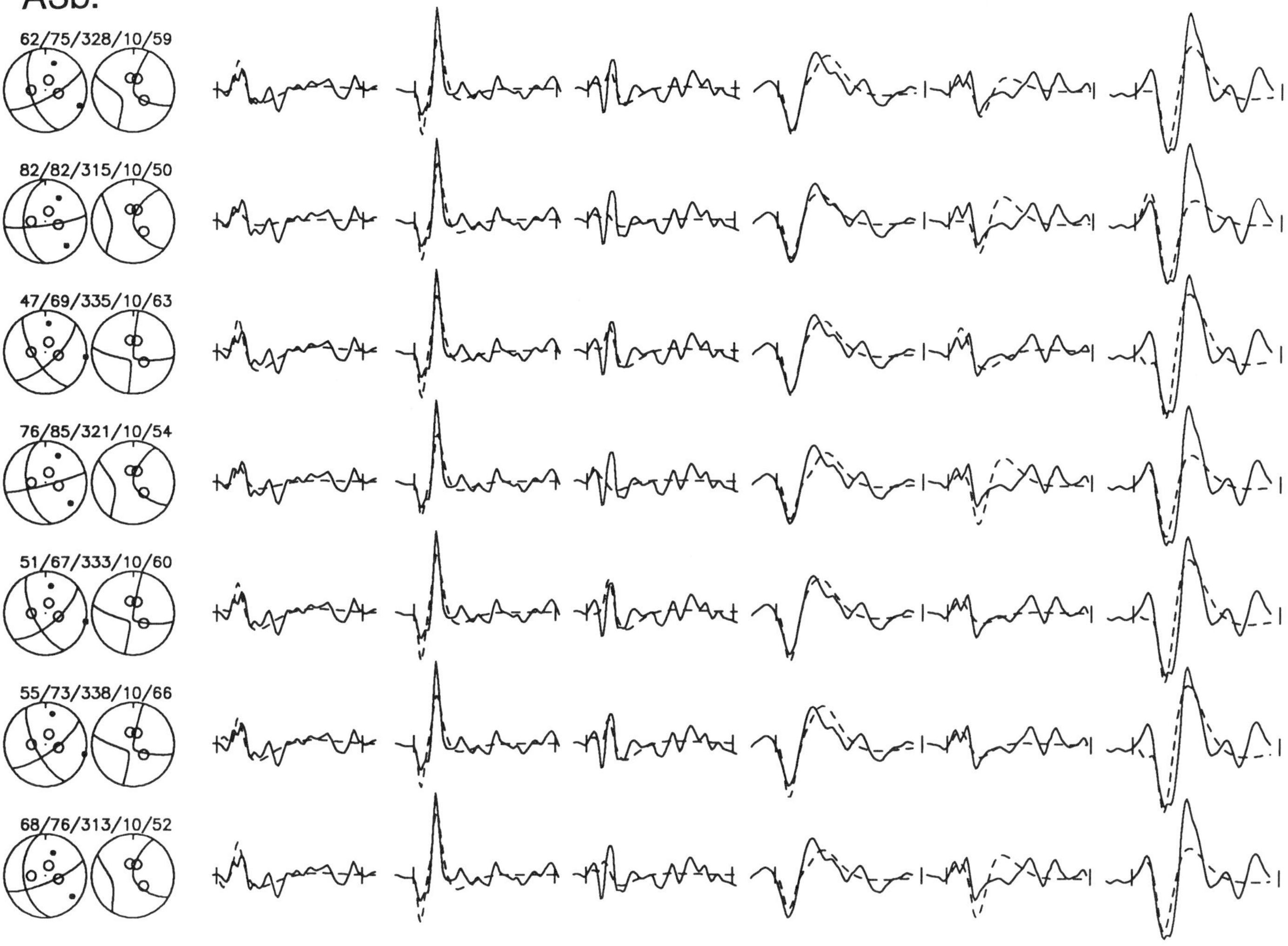

COP, KEV, GDH, COL, MAT, and SNG are a little too big, but again the synthetic SH phases at SNG and AAE fit well, better than for the "best" solution. In any case, although this solution is not as well constrained as others, there is no escaping its being different from those of the mainshock or the aftershock that occurred an hour before it.

1974 August 27 (Fig. A4). 39.52°N, 73.82°E, depth = 7 ± 3 km, strike = 5° ± 15°, dip = 50° + 20°/- 15°, rake = 341° + 10°/- 12°.

The fault plane solution of this aftershock differs from that of the mainshock and those of the two aftershocks discussed above (Balakina, 1983) (Fig. A4a). If the northerly trending plane is the fault plane, then a large component of strike-slip faulting occurred, as Jackson and others (1979) inferred from first motions of P waves. Nevertheless, a component of normal faulting, reflecting northeast-southwest extension, seems also to have occurred. The small amplitudes of the signals and the poor fits at some stations make the solution less well constrained than most of the others, but both the strike-slip and the normal components seem to be required.

A strike of 020°, 15° more counterclockwise than the "best"-fitting value, yields calculated P-wave amplitudes at HLW and JER that are a little too small, calculated SH amplitudes at COP, KEV, and CHG that are too big, but somewhat better fits of SH at AAE and NUR (Fig. A4b). A strike of 350° generates P-wave amplitudes at ATU and COP and SH amplitudes at AAE and CHG that are too big. The dip is particularly poorly constrained. For a relatively steep dip of 70°, 20° steeper than the "best"-fitting value, the calculated P-wave polarity at MAT is wrong, and SH at KEV is too small, but other waveforms are fit well. For a dip of 35°, the synthetic P phase at MAT is a little too big, whereas the synthetic P phases at HLW and JER are a little too small, and synthetic phases at CHG, AAE, and KEV are also too big. A gently plunging rake of 351° yields incorrect P-wave polarities at HLW and JER, and the synthetic P phase at AAE and the SH phase at KEV are too big. A relatively steep rake of 349°, however, seemed acceptable to us. A rake of 329° generates P phases at AAE, HLW, and MAT and an SH phase at ATU that are too large.

1978 October 8 (Figure A5). 39.39°N, 74.72°E, depth = 18 ± 5 km, strike = 114° + 25°/ - 20°, dip = 57° ± 10°, rake = 127° ± 10°.

This earthquake occurred about 35 km west of the much larger Wuqia earthquake of 1985 August 23, discussed below. The solution, constrained by both P-wave first motions and by waveforms, includes large components of both strike-slip and thrust faulting, on planes striking roughly northeast and northwest (Fig. A5a), but these planes dip in the opposite directions from those for the 1985 Wuqia earthquake. Hence, the same faults could not have been active. The large component of thrust faulting is well constrained by the compressional first motions at most stations (see also Balakina [1983]), and the strike-slip component is constrained largely by SH waveforms. Because of the rather large depth of 18 km, the direct SH and the reflected phase sSH arrive as distinct phases separate from one another, and their relative importance in constraining the solution can be assessed.

Not all phases can be matched with a common set of source parameters, and in particular, the direct SH phases at AQU and JER are too small for what we consider to be the "best"-fitting source parameters. Nevertheless, the source parameters are rather tightly constrained (Fig. A5b). A more southeasterly strike (139°) of the northwest-southeast–striking plane yields an improved fit of SH at JER, but not at AQU. Moreover, the synthetic direct SH phases at NAI and BAG for it are too small. For a more easterly strike (094°), not only are the synthetic P phases too big at AAE and in Europe (e.g., KBS) but also synthetic SH phases at AAE, JER, and MAT are out of phase with those observed. A relatively steep dip of 67° puts the P-wave nodal plane too close to the ray to AAE and an SH nodal surface too close to MAT. For a gentle dip of 47°, the synthetic P at AAE is too big, the synthetic SH phase at AAE is too small, and those at JER and MAT are out of phase. For a relatively gentle rake of 137°, corresponding to a relatively large strike-slip component, the sSH at AAE is a little too large, but the fits otherwise are not bad. For a relatively steep rake of 117°, corresponding to a large thrust component, the synthetic P phases at AAE and KBS are too big, and the synthetic SH at MAT is a little too small. The degradation in fits increase rapidly for rakes that deviate from 127° by more than 10°. Ekström's (1987) parameters lie well within the range given here, and so do those of Balakina (1983) based on P-wave first motions listed in the Bulletins of the International Seismological Centre.

1978 November 1 (Fig. A6). 39.35°N, 72.66°E, depth = 5 ± 3 km, if one event (a), strike = 111°, dip = 52, rake = 152, or if three sub-events (b), depths = 8 km, 9 km, and 10 km, strikes = 70°, 108°, and 116°, dips = 85°, 68°, and 50°, and rakes = 180°, 163°, and 150°.

Like those from the 1974 August 11 Markansu Valley earthquake, P phases begin with a small precursor. The first motions of the precursor call for large strike-slip components on planes striking north-northwest or east-northeast (Balakina, 1983), but waveforms cannot be fit by such a solution. When only one subevent was used, the inversion routine consistently returned a solution with oblique thrust faulting. A slight improvement was obtained by including rupture propagation, with a velocity of 2.5 km/s toward 315° yielding the most improvement, as shown in Figure A6a. Nevertheless, neither the small precursors nor later phases are fit well.

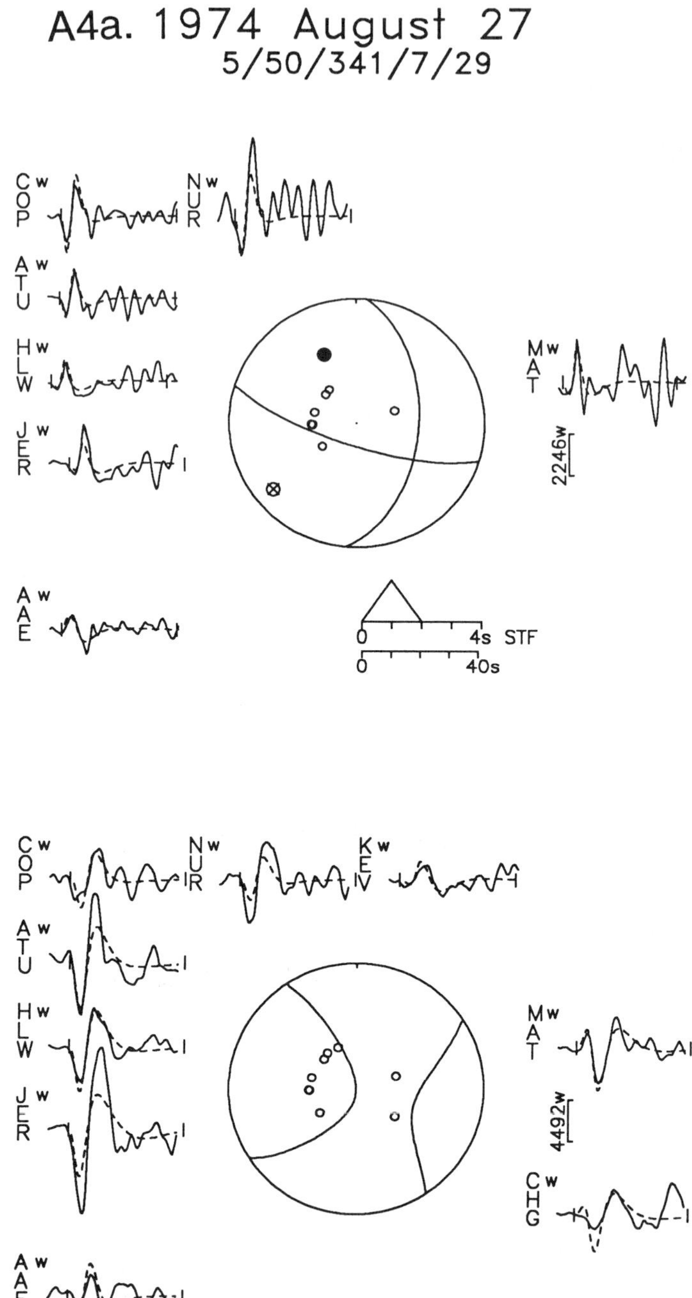

We then experimented with multiple events. Fixing a first subevent to match the small P-wave first motions did not help much,

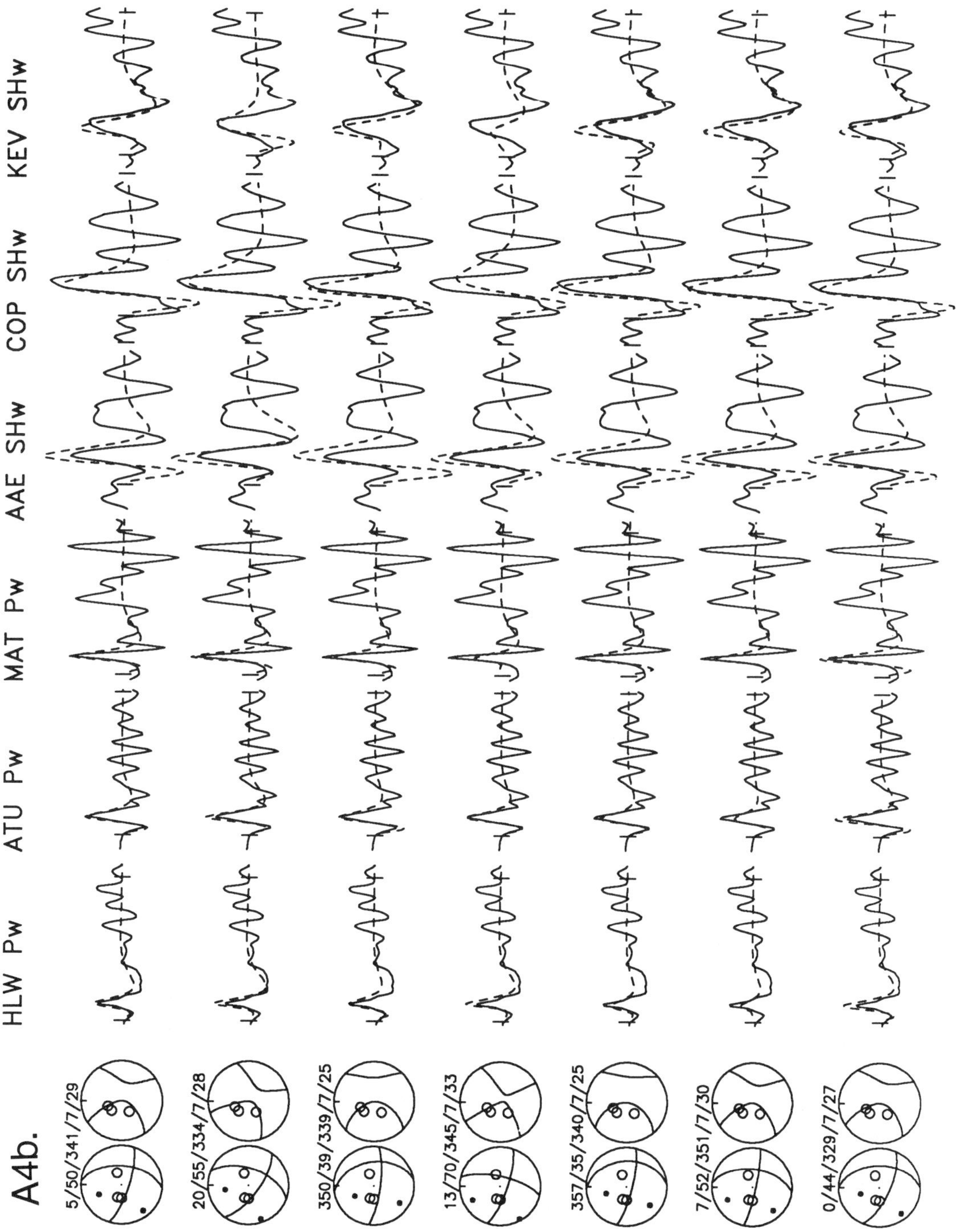
A4b.
HLW Pw
ATU Pw
MAT Pw
AAE SHw
COP SHw
KEV SHw
5/50/341/7/29
20/55/334/7/28
350/39/339/7/25
13/70/345/7/33
357/35/340/7/25
7/52/351/7/30
0/44/329/7/27

because the seismic moment for this subevent is very small. Including only one subevent after this initial small subevent also did not yield a very good fit. Finally, we experimented with three subevents, eventually allowing the fault solutions and the relative locations of the larger, later two subevents be free. In Figure A6b, the resulting matches of synthetic and recorded waveforms are shown. The most noticeable improvements are P phases at AAE and NAI, which are still not fit well, and SH at MAT and SHK. The conclusion, after much frustration, is that oblique thrust faulting occurred with shortening aligned north-northwest and with right-lateral slip on the southerly dipping plane. This inference is consistent with minor surface deformation in the source area of the earthquake (Nikonov and others, 1983).

1982 September 29 (Fig. A7). 37.35°N, 72.94°E, depth = 10 km, strike = 000° (± 20°), dip = 60° (± 10°), and rake = 290° (± 20°).

This earthquake was too small to be studied by synthesizing long-period waveforms from the WWSSN. Using digitally recorded signals, however, Ekström (1987) determined the focal depth and a fault plane solution showing largely normal faulting on northerly trending planes. The P-wave first motions corroborate the inference of normal faulting, and using them we altered his solution slightly.

1983 December 16 (Fig. A8). 39.33°N, 72.91°E, depth = 11 ± 4 km, strike = 192° + 20°/- 25°, dip = 53° ± 10°, rake = 58° ± 20°.

Despite the large number of stations that recorded this event, the solution is not well constrained (Fig. A8a). There is little doubt that a large component of thrust faulting is required, but the orientation is poorly resolved, as is clear from the uncertainties in the strike and rake (Fig. A8b). A 20° more clockwise strike (212°) renders synthetic P phases at JER and in Europe (e.g., KEV) too big, and SH phases in Europe too small. A 25° more counterclockwise strike (167°) gives synthetic P phases at KEV and GDH and SH phases at NWA0 and in Europe (JER, ATU, TRI, and COP) that are too big. A 10° steeper dip (63°) yields a synthetic P phase at JER that is too small and SH phases at NWA0 and COP that are too big. A 10° gentler dip (43°) generates synthetic P phases that are too big at JER, KEV, and GDH, and SH phases at COP, KEV, and GDH that are too small. A rake (78°) larger and steeper by 20° than the "best"-fitting value yields a synthetic P phase at GDH that is a little too big, an SH at JER that is too small, and sSH pulses at ATU, TRI, and COP that are too small. A rake (38°) smaller by 20° generates a synthetic P phase at BAG that is too small and SH pulses at NWA0, ATU, and TRI that are too big.

1984 October 26 (Fig. A9). 39.18°N, 71.34°E, depth = 9 ± 3 km, strike = 68° + 25°/- 20°, dip = 44° +9°/- 6°, rake = 57° ± 12°.

The predominantly compressional P-wave first motions require a large component of thrust faulting, but the SH phases indicate a small right-lateral strike-slip component on the plane dipping south (Fig. A9a).

The uncertainties are constrained largely by SH phases (Fig. A9b). The strike is not well constrained, and a large difference from the "best"-fitting value is permitted by the match of synthetic with recorded seismograms. For a 25° more clockwise strike of 93°, the synthetic SH phase at COL is a little too large, and those phases at TOL, KONO, and NUR are a little too small. For a 20° more counterclockwise strike of 48°, the initial motions of the synthetic SH phases at NUR and COL are a little too small, and SH phases at TOL and KONO are a little too large. For a 9° steeper dip of 53°, the synthetic sSH phase at COL is too big, and the SH phases at KONO and

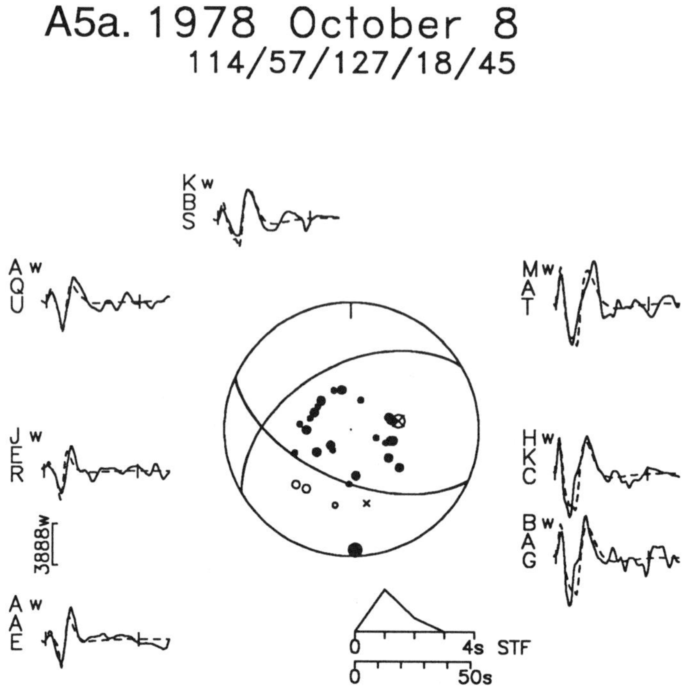

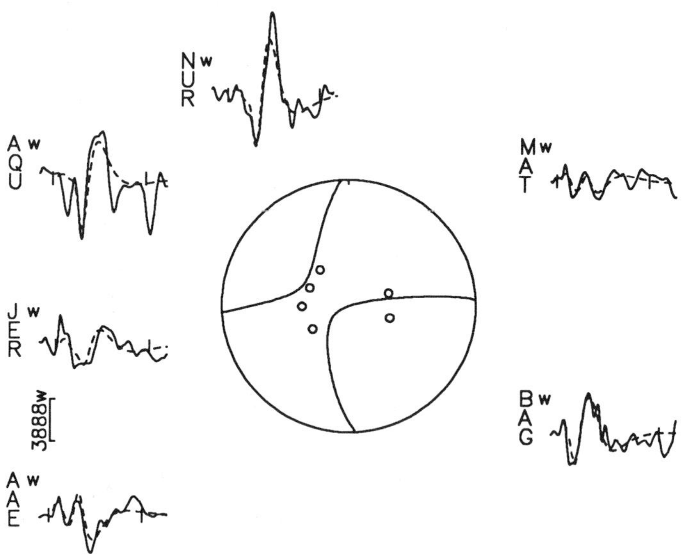

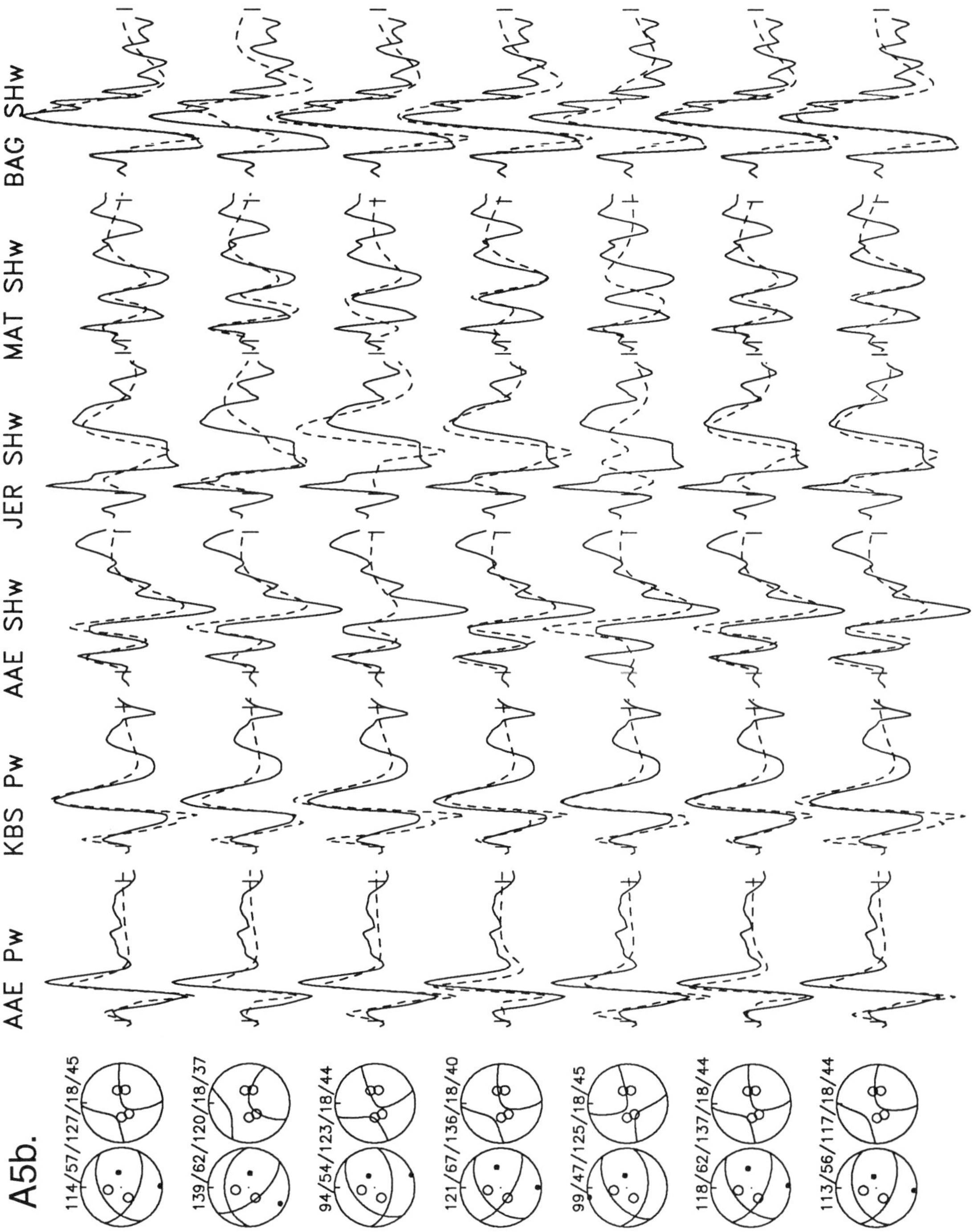
A5b.
AAE Pw
KBS Pw
AAE SHw
JER SHw
MAT SHw
BAG SHw
114/57/127/18/45
139/62/120/18/37
94/54/123/18/44
121/67/136/18/40
99/47/125/18/45
118/62/137/18/44
113/56/117/18/44

1978 November 1 (1 subevent)

111/52/152/5/976

A6a.

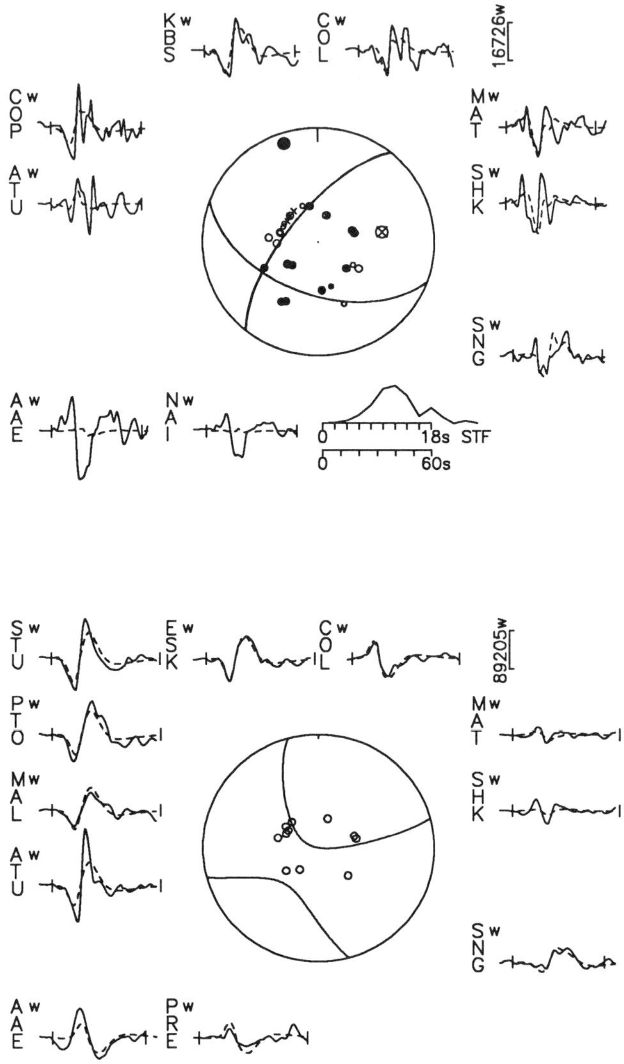

1978 November 1 (3 subevents)

1:70/85/180/8/78

2:108/68/163/9/426

3:116/50/150/10/436

A6b.

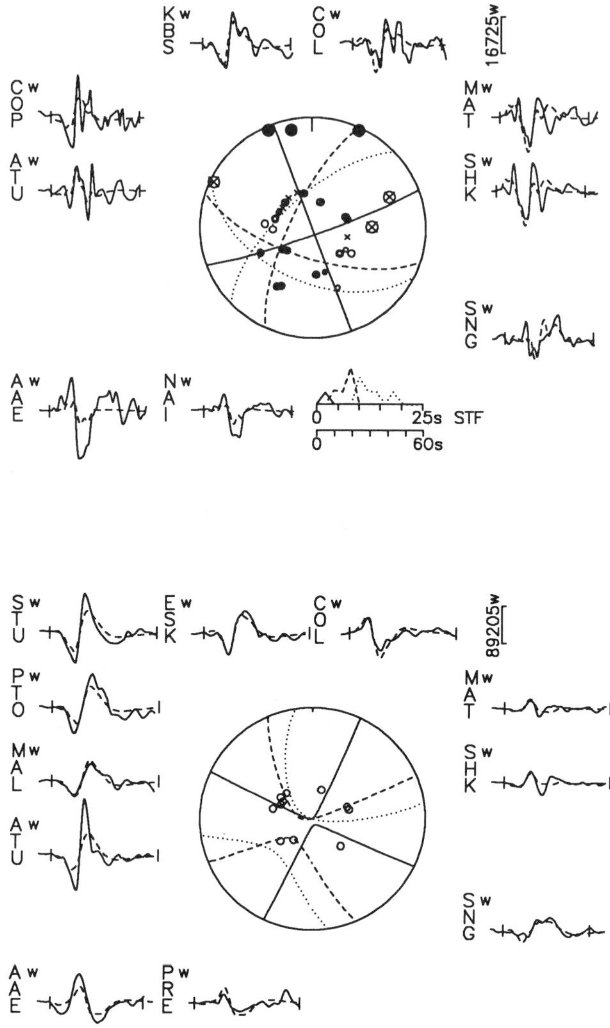

A7. 1982 September 29

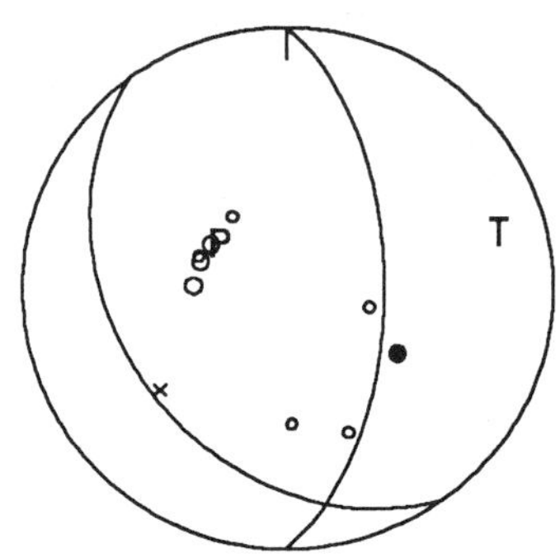

NUR are too small. For a 6° gentler dip of 38°, the synthetic SH phases are too big at COL, too small at NAI, and out of phase at TOL. For a 12° steeper rake of 69°, the synthetic P phase at COL is too big, and the synthetic SH phases at NAI, TOL, KONO, NUR, and GDH are a little too small. For a 12° gentler rake of 45°, the synthetic SH phases at TOL, ESK, KEV, and COL are too big.

1985 August 23 Wuqia earthquake (Fig. A10). 39.43°N, 75.27°E, depth = 18 km, with average solution of strike = 316°, dip = 46°, rake = 160°, $M_o = 2.22 \times 10^{19}$ Nm, or consisting of three subevents: strike = 283°, dip = 34°, rake = 142°, $M_o = 3.6 \times 10^{18}$ Nm; strike = 326°, dip = 58°, rake = 158°, $M_o = 7.9 \times 10^{18}$ Nm; and strike = 309°, dip = 41°, rake = 152°, $M_o = 9.9 \times 10^{18}$ Nm.

This is a large earthquake for which the complexity of the signals requires a source more complicated than a point source. We explored a series of possible sources that include a single point source (Fig. A10a), a source with a uniform rupture velocity (Fig. A10b), and multiple sources (Fig. A10c). Definite improvements in matches of synthetic and recorded waveforms can be seen with increasingly complex sources, described by an increasing number of free parameters, but as a result it is difficult to assign uncertainties to all of these parameters.

Note first that the strong P waves at stations to the northwest and the small complex signals at stations to the east and southeast require both that one nodal plane pass near the stations to the east and southeast and that there be a significant strike-slip component. These inferences are corroborated by the amplitudes and shapes of SH phases. Neither the broad pulses for both P and SH to the east and southeast, however, nor the multiple pulses of P at these stations and of SH at stations to the northwest are synthesized well using a point source (Fig. A10a).

After numerous trials and errors we examined systematically rupture propagation toward the northwest, similar to that inferred by Ekström (1987). In Figure A10b we show the matches of recorded and synthetic waveforms for a rupture velocity of 2.5 km/s in the direction 300°. With velocities of 2.0, 2.5, or 3.0 km/s in the direction 300°, the improvement over a point source is noteworthy, including broader synthetic signals to the east and southeast and a better match of the complex P and SH phases to the northwest and east, respectively. In all cases, however, the source time functions appeared to consist of three subevents. Consequently, we then broke the uniform rupture into separate subevents, each describable as point sources. Fixing their locations, we inverted for the source parameters of three subevents. which allowed a still better match to the complexities of P phases to the northwest and SH phases to the southeast. Finally, we inverted for the source parameters of these subevents, including not just strikes, dips, rakes, depths, and seismic moments but also their relative positions and times, to obtain the final fits shown in Figure A10c. The addition of these parameters also yielded improved fits to the complex P phases at stations to the west.

Obviously assigning uncertainties to each of the parameters is hopeless, and to place constraints on some of them, we considered only the point source. First, for an earthquake of this size and for which there are suggestions of surface faulting (Feng and others, 1986), the depth is not a very meaningful quantity. The inversion routine repeatedly returned depths of about 18 km, and we simply accept that the depth was shallower than 25 km but probably not confined to the upper 10 km.

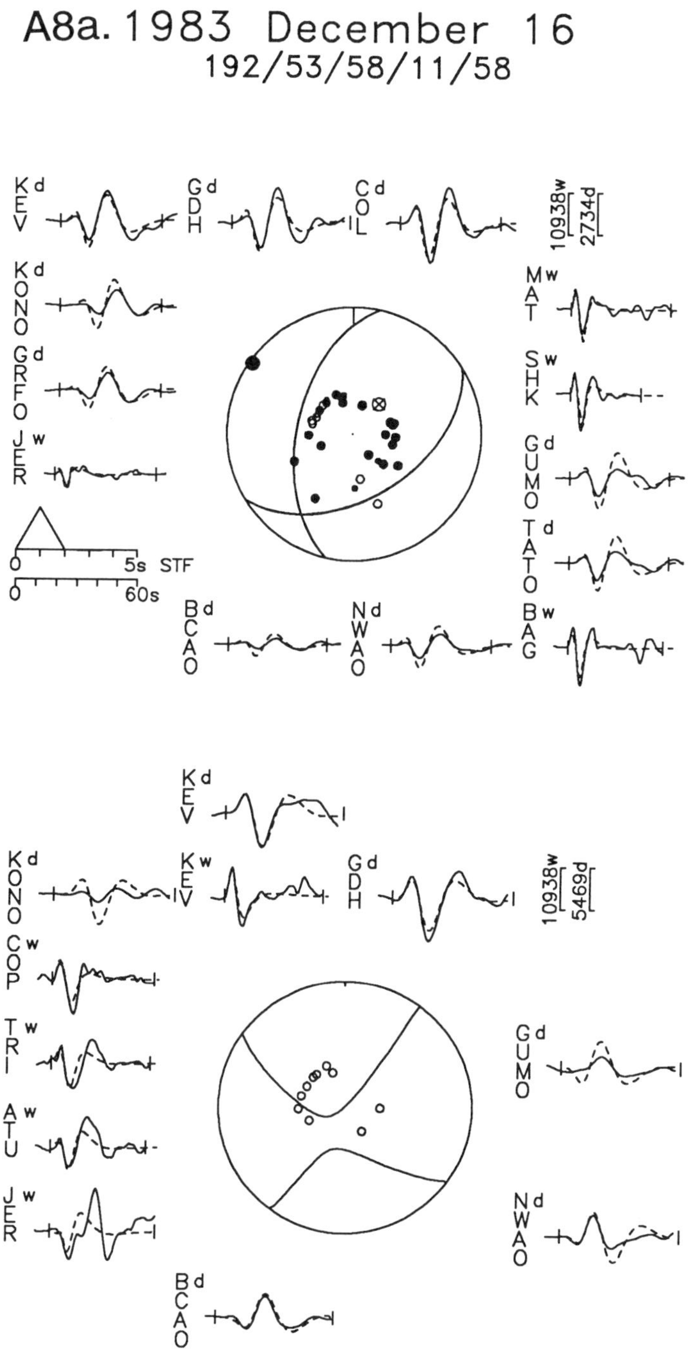

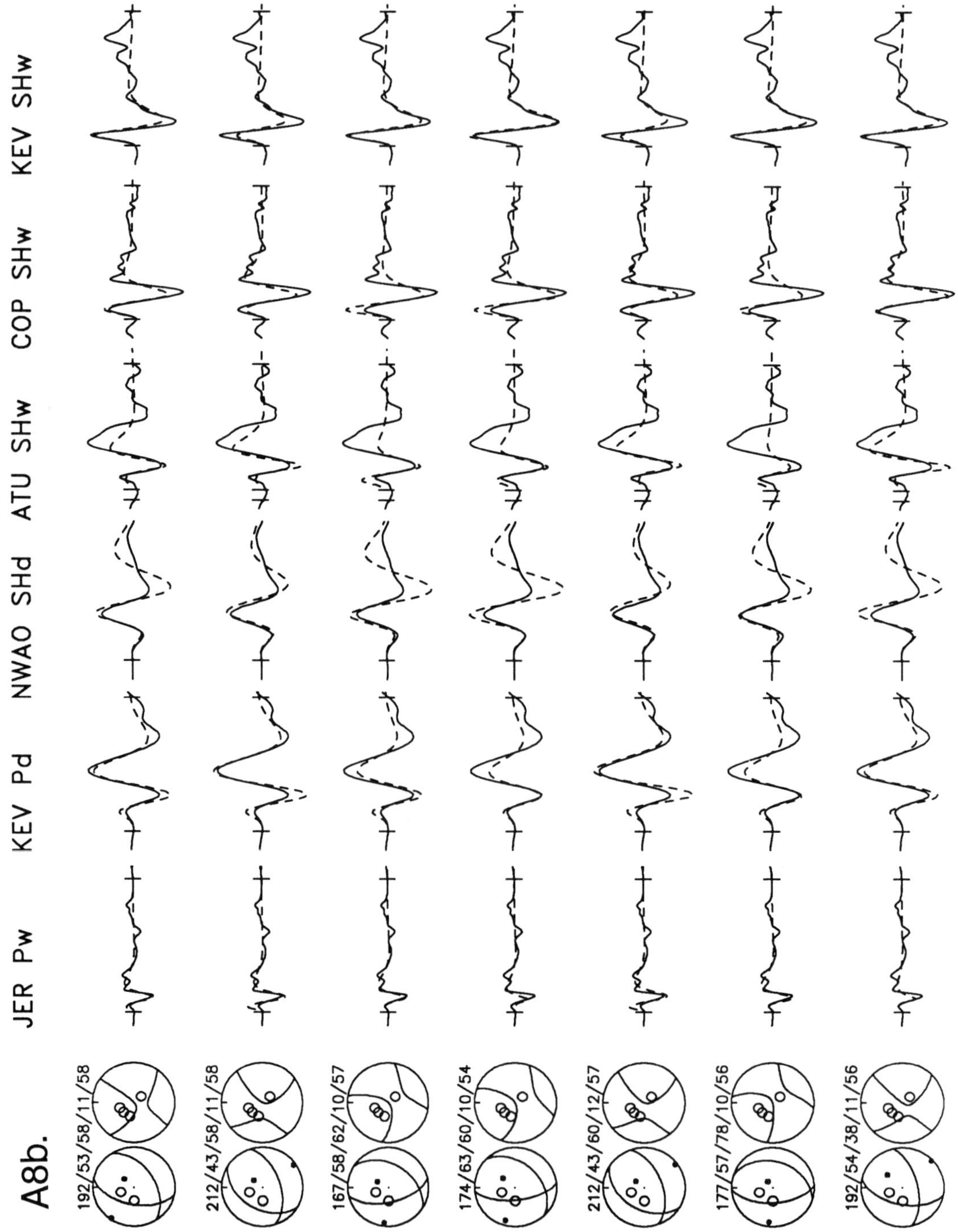
A8b.
JER Pw
KEV Pd
NWAO SHd
ATU SHw
COP SHw
KEV SHw
192/53/58/11/58
212/43/58/11/58
167/58/62/10/57
174/63/60/10/54
212/43/60/12/57
177/57/78/10/56
192/54/38/11/56

Let us consider the degradation of fits of synthetic and recorded waveforms with systematic variations in the strike, dip, and rake of the northeast-striking plane, which is probably the auxiliary plane. To do this, we used the match with a rupture propagating at 2.5 km/s in the direction N300° to define an acceptably good fit. A more clockwise strike of 85° renders synthetic P waveforms at MUN and LEM too big, pulse widths of synthetic SH waveforms at SNG and LEM too sharp, and SH waveforms at European stations (e.g., VAL) poor. For the more counterclockwise value of 35°, synthetic P waveforms at STU, VAL, ESK, AKU, GDH, and COL fit poorly, and the polarities of synthetic SH phases at SNG and LEM are opposite to those observed. For a steep dip of 78°, synthetic P phases at STU, GDH, and SEO and SH phases at VAL, ESK, and AKU are a little too small. For a gentle dip of 58°, synthetic P phases at MUN and LEM are too big and SH phases at VAL, ESK, AKU, and GUA are too small. For a relatively steep rake of 57°, synthetic SH phases at SEO and SHK are a little too big, and those at VAL, ESK, and AKU are too small. For a relatively gentle value of 32°, the synthetic P waveforms at STU, VAL, GDH, and COL are too small, and SH phases at SNG, LEM, and MUN are a little too big. These ranges of values encompass those given by Ekström (1987) for this earthquake.

1985 September 11 (Fig. A11). 39.36°N, 75.44°E, depth = 7 ± 3 km, strike = 108° + 30°/- 60°, dip = 31° + 7°/-10°, rake = 104° + 25°/-60°.

Obtaining an acceptable solution for this was especially frustrating. Only three P waves could be digitized, and all of them were recorded during noisy periods (Fig. A11a). Waveforms from DAV are notoriously inconsistent (e.g., Chen and others, 1981; Molnar and Lyon-Caen, 1989). Although we plot the recorded and synthetic waveforms from TRI and BAG, we did not rely on them much in constraining the nodal planes. The signal at TRI is superimposed on the coda from another earthquake. Moreover, broad-band signals constructed from digital recordings by Ekström (1987) exhibit similar low signal-to-noise ratios. Because only a couple of other reliable P-wave first motions could be identified with confidence, the fault plane solution is not well constrained by them. The source parameters are constrained largely by SH phases, and they fit two different classes of solution: largely strike-slip displacement on northwesterly or northeasterly trending planes or largely thrust faulting on east-west–trending planes. Ekström (1987) reported the latter type, which we favor, but not with much confidence.

In any case, the SH phases could not be fit as well as we hoped with either type of solution. The double pulse at some European stations (e.g., COP and NUR) has eluded a simple explanation. In addition, our calculated moment is roughly 3 times smaller than Ekström's (1987), which in part is due to his much longer time function but also reflects a difficulty in getting a good fit. Finally, travel time residuals for these phases are also quite large, commonly several seconds, suggesting that either the phases have been misidentified or that the location and origin time are in error. We spent considerable time redigitizing and realigning the recorded waveforms with their synthetic counterparts generated for different fault plane solutions but with no significant improvement in matches.

Notice that a strike of 138°, 30° more clockwise than the "best" value, yields a yet worse fit of P at TRI and SH phases at COL and BAG that are too small and too big, respectively (Fig. A11b). For a strike of 48°, 60° counterclockwise from the "best" strike, P at TRI is

A9a. 1984 October 26
68/44/57/9/145

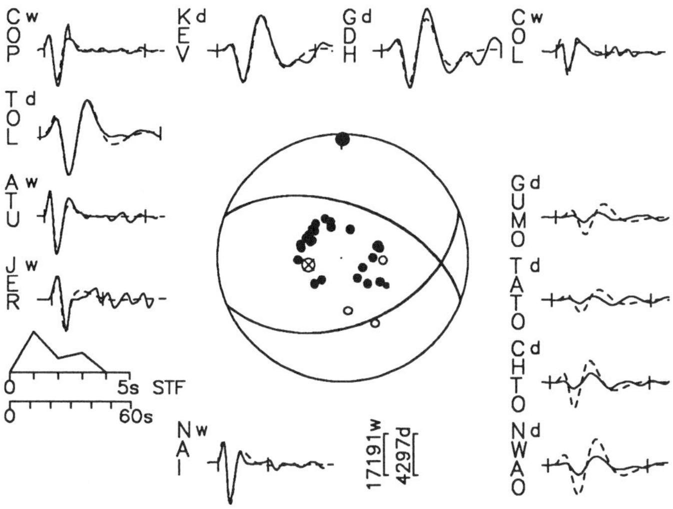

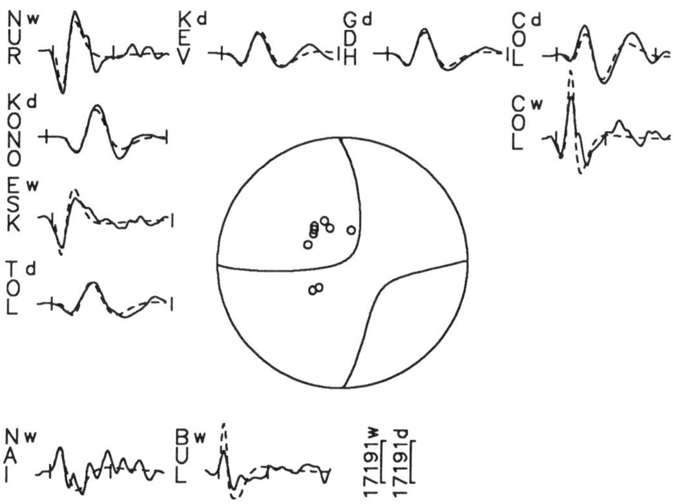

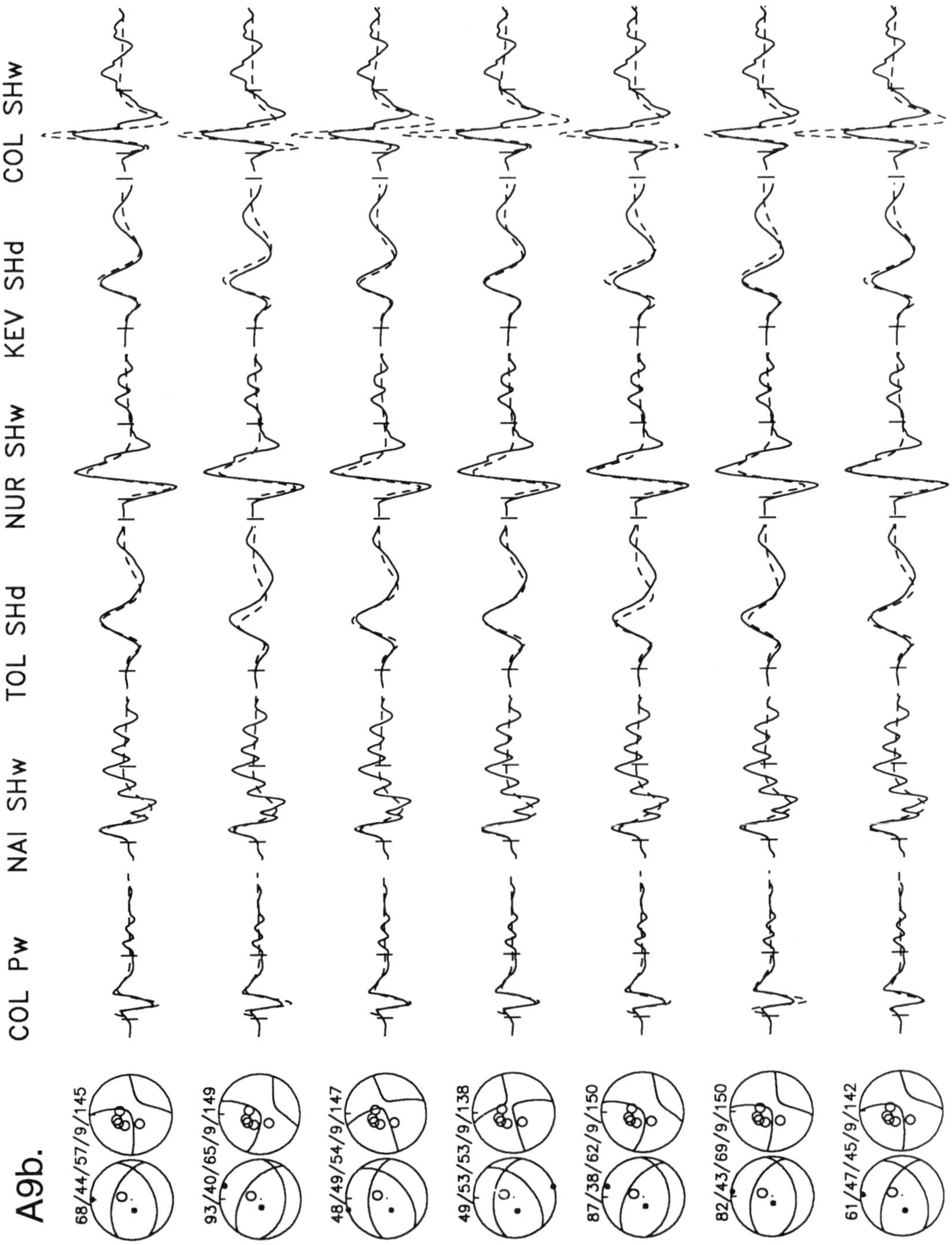
A9b.
COL Pw
NAI SHw
TOL SHd
NUR SHw
KEV SHd
COL SHw
68/44/57/9/145
93/40/65/9/149
48/49/54/9/147
49/53/53/9/138
87/38/62/9/150
82/43/69/9/150
61/47/45/9/142

A10a. 1985 August 23
309/46/152/18/2030

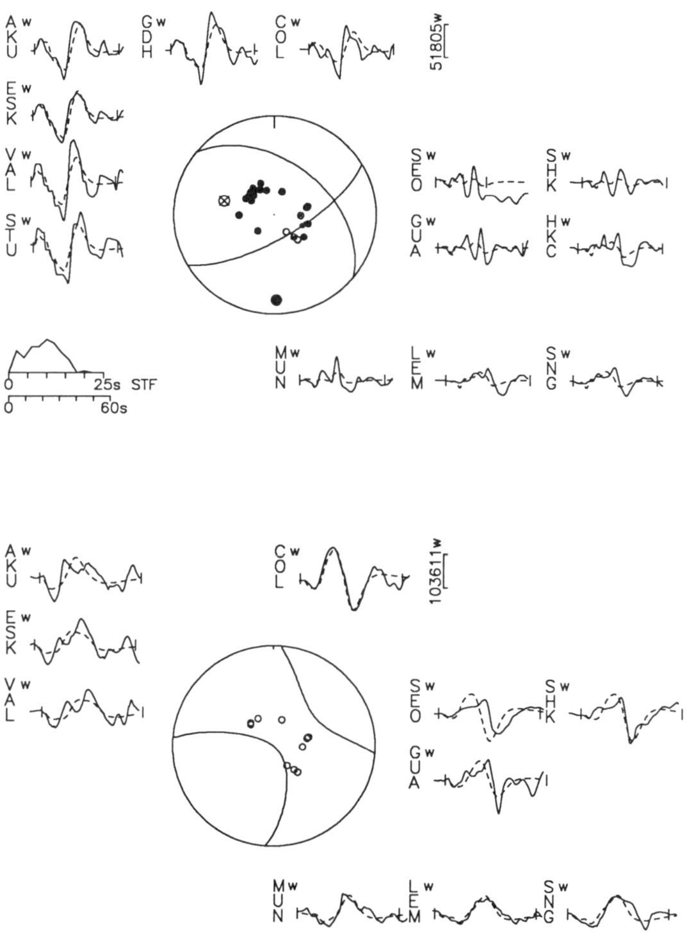

also worse, but at BAG it is better than for the "best," and the synthetic SH at ATU is too big. For a steeper dip of 38°, synthetic P phases at TRI and BAG are too big, and SH phases at ATU, STU, and BAG are too small. For a gentler dip of 21°, the synthetic P at TRI again is too small, but that at BAG fits well, ad SH phases at COL ad BAG are a little too large. For a rake of 129°, 25° larger than the "best," the synthetic P at TRI is too small, the synthetic SH at STU is a little too small, and those COL and BAG are too big. For a rake of only 44°, 60° smaller than the "best," synthetic SH phases at ATU and STU are too big, and that at COL is too small.

1985 October 13 (Fig. A12). 40.32°N, 69.84°E, depth = 15 ± 3 km, strike = 268° + 10°/- 35°, dip = 50° + 7°/- 5°, rake = 77° + 12°/- 8°.

P-wave first motions require a large thrust component but do not place a tight constraint on the orientations of the planes (Fig. A12a). The match of synthetic to recorded seismograms requires that both planes strike roughly east-west to perhaps northeast-southwest.

The strike is not very well constrained (Fig. A12b). A 10° more clockwise strike of 278° yields a synthetic P phase at TRI that is a little too big and synthetic SH phases at COL and TRI that are too small. A 35° more counterclockwise strike of 233°, corresponding also to a 12°

A10b. 1985 August 23
316/46/160/18/2225

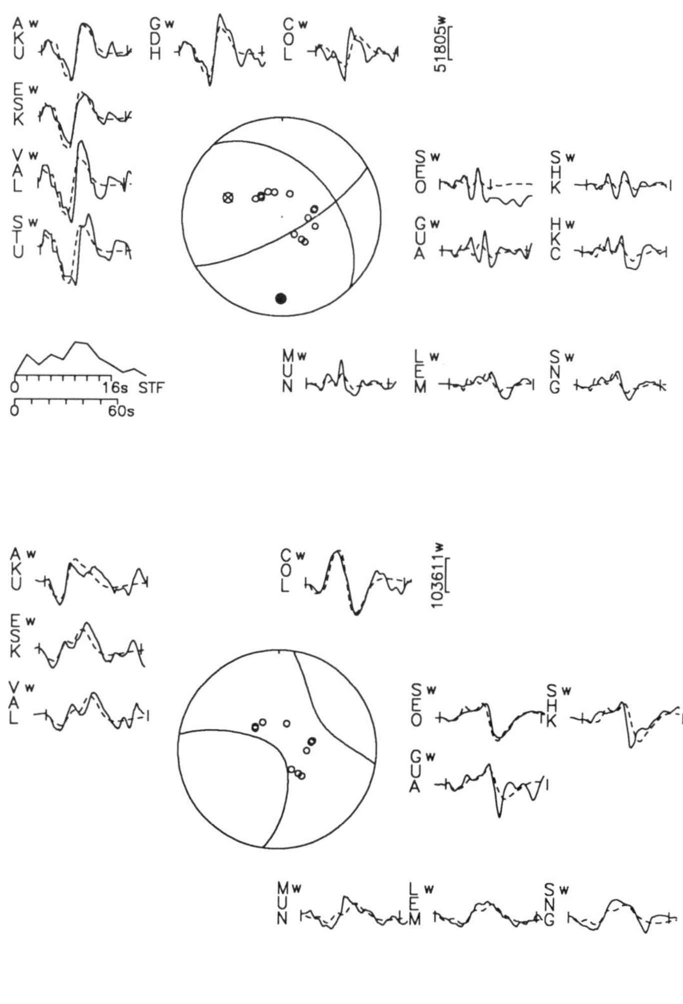

A10c. 1985 August 23

1:282/34/142/18/359
2:326/58/158/19/791
3:309/41/152/18/992

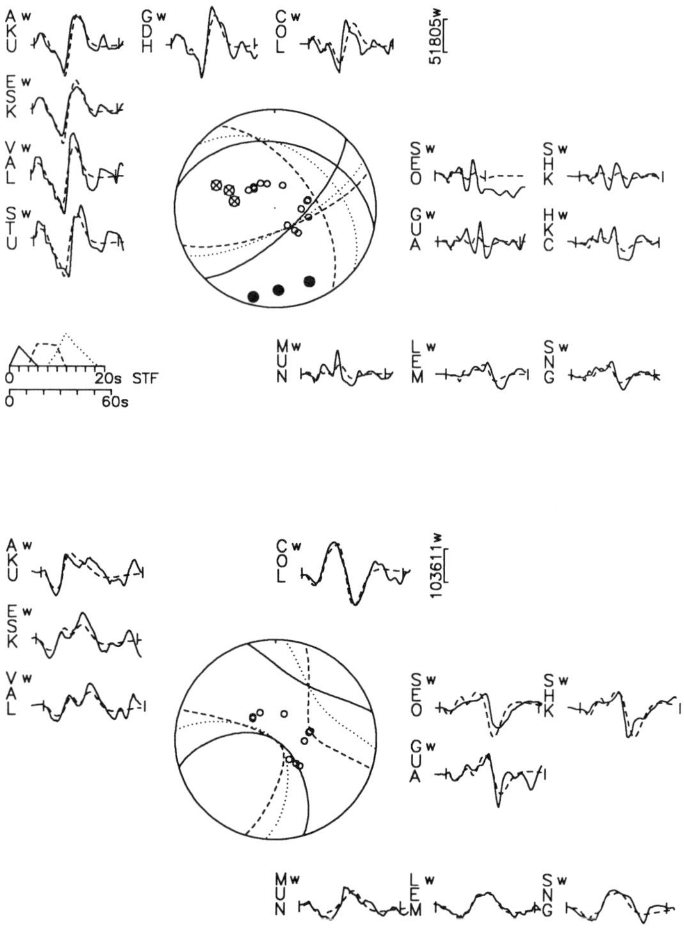

A11a. 1985 September 11

108/31/104/7/81

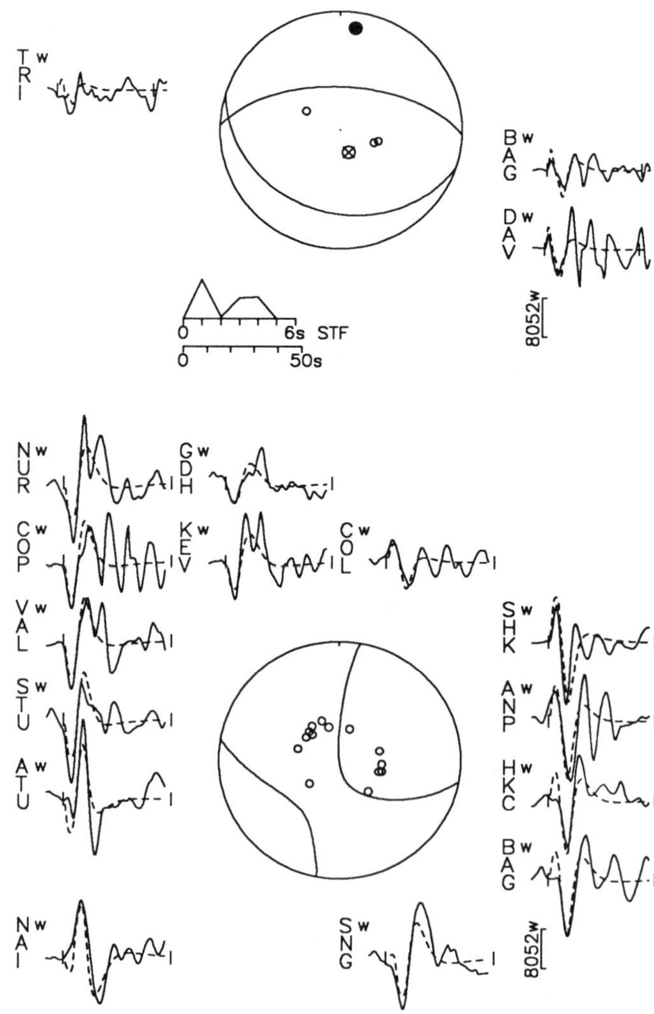

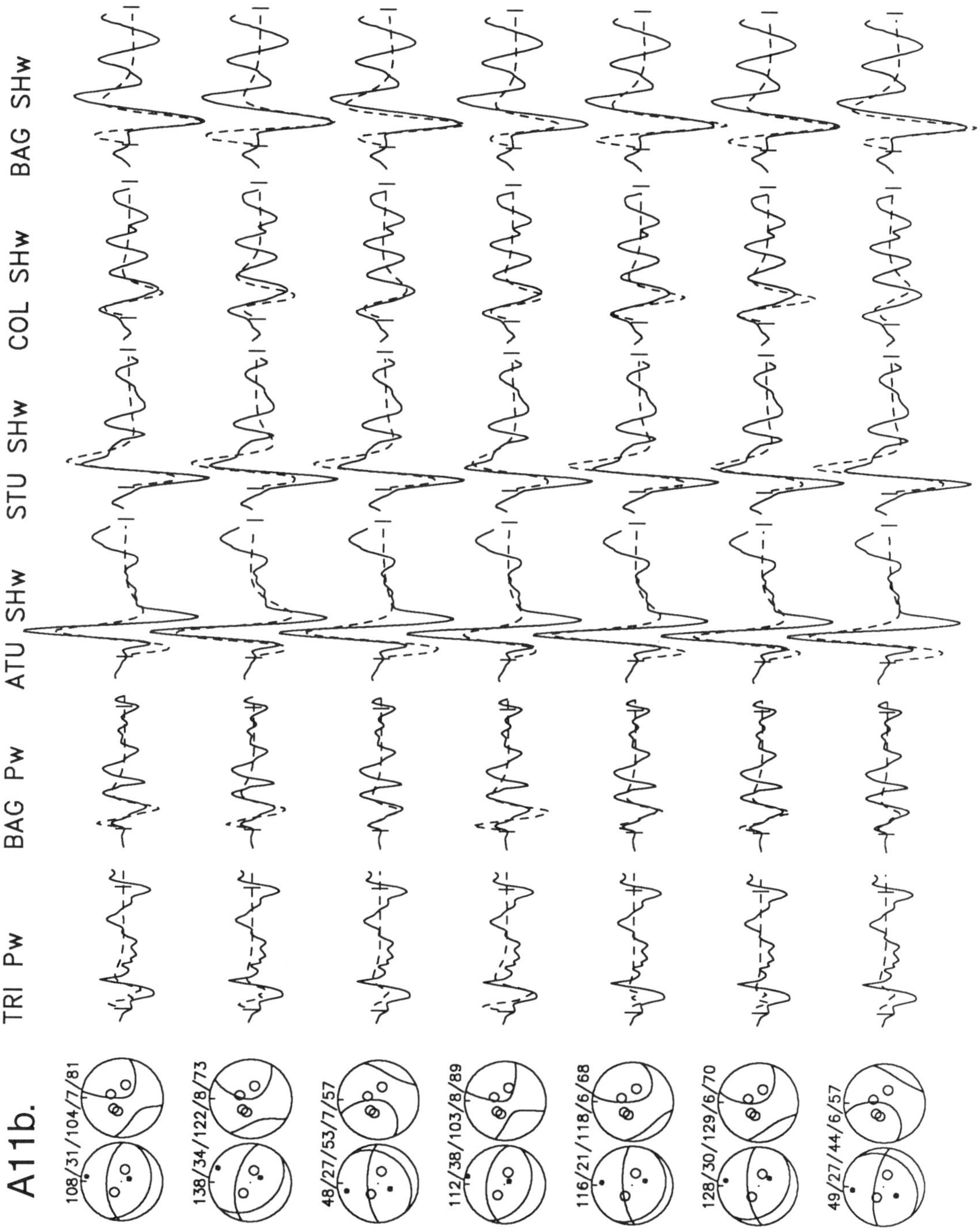
A11b.
TRI Pw
BAG Pw
ATU SHw
STU SHw
COL SHw
BAG SHw
108/31/104/7/81
138/34/122/8/73
48/27/53/7/57
112/38/103/8/89
116/21/118/6/68
128/30/129/6/70
49/27/44/6/57

A12a. 1985 October 13
268/50/77/15/40

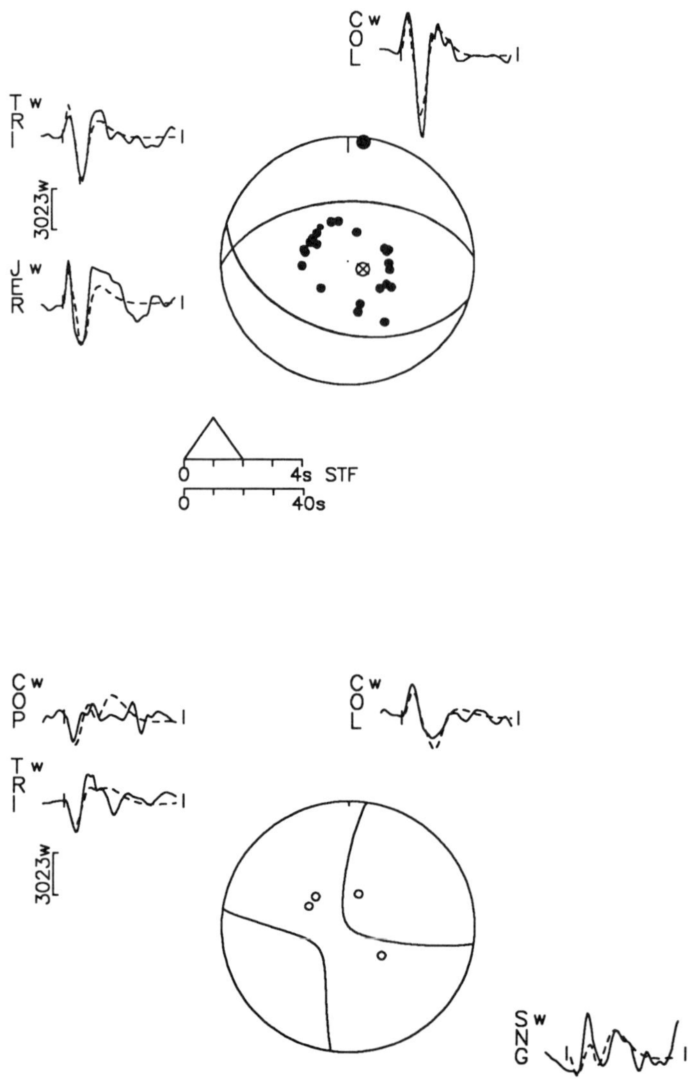

steeper rake of 89°, yields a synthetic SH phase at COP that is too small. A 7° steeper dip of 57° generates a P phase at COL that is too small and a sSH that is too large. A 5° gentler dip makes a P phase at TRI that is a little too big and SH phases at COP, TRI, and COL that are too small. An 8° gentler rake of 69° creases SH phases at COL and TRI that are too small, but also with a large backswing at COL.

REFERENCES CITED

Abers, G., Bryan, C., Roecker, S., and McCaffrey, R., 1988, Thrusting of the Hindu Kush over the southeastern Tadjik Basin, Afghanistan: Evidence from two large earthquakes: Tectonics, v. 7, p. 41–56.

Andrews-Speed, C. P., and Brookfield, M. E., 1982, Middle Paleozoic to Cenozoic geology and tectonic evolution of the northwestern Himalaya: Tectonophysics, v. 82, p. 253–275.

Argand, E., 1924, La Tectonique de l'Asie: International Geological Congress Reports of Sessions, v. 13, p. 170–372.

Artemijev, M. E., and Belousov, T. P., 1983, Recent tectonics and isostasy in the Pamirs, southern Tien Shan and the zone of their junction: Bolletin Geofisica Teoria Applicada, v. 25, p. 401–408.

Baker, D. M., Lillie, R. J., Yeats, R. S., Johnson, G. D., Yousuf, M., and Zamin, A.S.H., 1988, Development of the Himalayan frontal thrust zone: Salt Range, Pakistan: Geology, v. 16, p. 3–7.

Balakina, L. M.,1983, Tectonics and mechanisms of earthquakes (in Russian), *in* Gubin, I. Ye., and Zakharov, S. A., eds., Tectonics of the Tien Shan and Pamir: Moscow, Nauka, p. 24–39.

Baranowski, J., Armbruster, J., Seeber, L., and Molnar, P., 1984, Focal depths and fault plane solutions of earthquakes and active tectonics of the Himalaya: Journal of Geophysical Research, v. 89, p. 6918–6928.

Baratov, R. B., ed., 1976, Separation of stratigraphic and intrusive formations of Tadjikistan (in Russian): Dushanbe, Donish, 268 p.

Bassoulet, J. P., Boulin, J., Colchen, M., Marcoux, J., Mascle, G., and Montenat, C., 1980, L'évolution des domaines téthysiennes au pourtour du bouclier indien du Carboniférе au Cretacé, *in* Géologie des chaines alpines issues de la Tethys: Bureau Recherche Géologie et Miniérs, Mèmoire, Orleans, France, Mèmoire 115, p. 180–198.

Bazhenov, M. L., and Burtman, V. S., 1981, Formation of the Pamir-Punjab syntaxis: Implications from paleomagnetic investigations of Lower Cretaceous and Paleogene rocks of the Pamirs, *in* Contemporary scientific researches in Himalaya: Dehra Dun, India, Bishen Singh Mahendra Pal Sing, p. 71–81.

——, 1982, The kinematics of Pamir arc: Geotectonics, v. 16 (English translation), p. 288–301.

——, 1986, Tectonics and paleomagnetism of structural arcs of the Pamir-Punjab syntaxis: Journal of Geodynamics, v. 5, p. 383–396.

——, 1990 Structural arcs of the Alpine Belt: Carpathians-Caucasus-Pamir (in Russian): Moscow, Nauka, 168 p.

Bekker, Ya. A., Koshlakov, G. V., and Kuznetsov, Ye. S., 1974a, Deep structure of southwest Tadjikistan from geologic and geophysical data (in Russian), *in* Searches for precursors of earthquakes in prediction polygons: Moscow, Nauka, p. 16–23.

Bekker, Ya. A., Koshlakov, G. V., Kuznetsov, E. C., Murchaidze, D. P., and Orlov, E. S., 1974b, Toward the tectonics of the region of the city of Dushanbe (Gissar Valley) from the results of geological-geophysical investigations (in Russian), *in* Searches for precursors of earthquakes in prediction polygons: Moscow, Nauka, p. 24–29.

Bekker, Ya. A., Konovalov, Yu. F., Koshlakov, G. V., and Muchaidze, D. R., 1983, New data about the structure of the earth's crust of Tadjikistan (in Russian), *in* Gubin, I. Ye., and Zakharov, S. A., eds., Tectonics of the Tien Shan and Pamir: Moscow, Nauka, p. 118–123.

Belokopytov, V. A., Guseva, T. V., Lukk, A. A., Skovorodkin, Yu. P., Trapeznikov, Yu. A., and Shevchenko, V. I., 1992, Present-day geodynamics and seismicity in the Garm test area, Tadjikistan: Tectonophysics, v. 202, p. 163–167.

Belousov, T. P., 1976, Tectonic movements of the Pamir in the Pleistocene and Holocene epochs and seismicity (in Russian): Moscow, Nauka, 120 p.

Beloussov, V. V., and 11 others, 1980, Structure of the lithosphere along deep seismic sounding profile: Tien Shan–Pamirs–Karakorum–Himalayas:

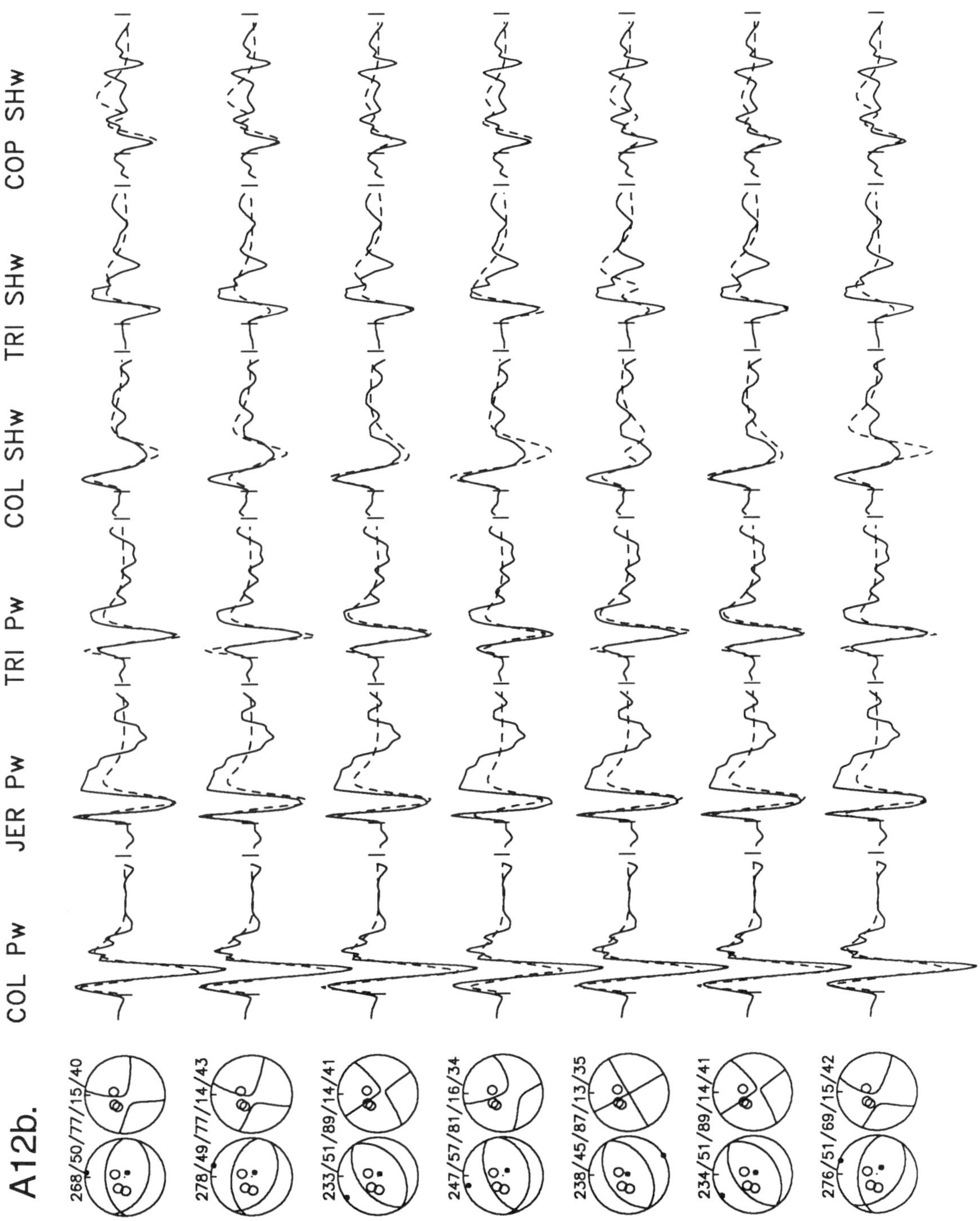
A12b.
COL Pw
JER Pw
TRI Pw
COL SHw
TRI SHw
COP SHw
268/50/77/15/40
278/49/77/14/43
233/51/89/14/41
247/57/81/16/34
238/45/87/13/35
234/51/89/14/41
276/51/69/15/42

Tectonophysics, v. 70, p. 193–221.

Belov, V. I., Kafarskii, A. Kh., and Pashkov, B. R., 1982, Toward a stratigraphy of the Paleozoic deposits in the Darvaz-Sarykol' zone of the Northern Pamir (in Russian): Bulletin MOIP (of the Moscow Society for the Investigation of Nature), v. 57, p. 33–45.

Bian Qiantao, 1992, On the tectonic characteristics and evolution of the Hoh Xil region, Qinghai Province: International Symposium on the Karakorum and Kunlun Mountains, June 5–9, 1992, Kashi [*sic*], China, Abstracts, p. 19.

Billington, S., Isacks, B. L., and Barazangi, M., 1977, Spatial distribution and focal mechanisms of mantle earthquakes in the Hindu-Kush–Pamir region: A contorted Benioff zone: Geology, v. 5, p. p. 699–704.

Blaise, J., Bordet, P., Montenat, C., Desparmet, R., and Marin, P., 1977, Recherches géologiques dans les montagnes centrales de l'Afghanistan (Hazarajat et sa bordure orientale), *in* Livre à la mémoire de Albert F. de Lapparent, consacré aux recherches géologiques dans les Chaines Alpines de l'Asie du Sud-ouest: Societé Géologique de la France Mémoire hors-série 8, p. 117–143.

Blaise, J., Bordet, P., Carbonnel, J.-P., and Montenat, C., 1978, Flyschs et ophiolites dans la région de Panjaw: Une suture néocimmérienne en Afghanistan central: Bulletin de la Societé Géologique de la France, v, 20, p. 795–798.

Boulin, J., 1981, Afghanistan structure, greater India concept and eastern Tethys evolution: Tectonophysics, v. 72, p. 261–287.

Boulin, J., and Bouyx, E., 1977, Introduction à la géologie de l'Hindou Kouch occidental, *in* Livre à la mémoire de Albert F. de Lapparent, consacré aux recherches géologiques dans les Chaines Alpines de l'Asie du Sud-ouest, Societé Géologique de la France Mémoire hors-série 8, p. 87–105.

Bratash, B. I., Yegupov, S. V., Pechnikov, V. V., and Shelomentsev, A. I., 1970, Geology and oil and gas in northern Afghanistan (in Russian): Moscow, Nedra, 288 p.

Brunel, M., Arnaud, N., Tapponnier, P., Vidal, Ph., Pan, Y., and Wang, Y., 1992a, North vergent nappe tectonics in the Pamir–Kun Lun: Preliminary Ar/Ar Jurassic age: International Symposium on the Karakorum and Kunlun Mountains, June 5–9, 1992, Kashi [*sic*], China, Abstracts, p. 28.

Brunel, M., and six others, 1992b, Tectonics in the eastern Pamirs: International Symposium on the Karakorum and Kunlun Mountains, June 5–9, 1992, Kashi [*sic*], China, Abstracts, p. 29.

Budanov, B. I., and Pashkov, B. R.,1988, On the scale of early Carboniferous and Permian vulcanism in the eastern part of the Northern Pamir (in Russian): Bulletin MOIP (of the Moscow Society for the Investigation of Nature), Geological Section, v. 63, p. 33–38.

Burov, E. B., Kogan, M. G., Lyon-Caen, H., and Molnar, P.,1990, Gravity anomalies, the deep structure, and dynamic processes beneath the Tien Shan: Earth and Planetary Science Letters, v. 96, p. 367–383.

Burtman, V. S., 1964, The Talaso-Fergana strike-slip fault (in Russian): Moscow, Nauka, Trudy Geological Institute, Akademi Nauk, USSR, 104, 143 p.

——, 1975, Structural geology of the Variscan Tien Shan, USSR: American Journal of Science, v. 272A, p. 157–186.

——, 1982, Development of the Pamir-Punjab syntaxis: Geotectonics (English translation), v. 16, p. 383–388.

Burtman, V. S., Peive, A. V., and Ruzhentsev, S. V., 1963, The main lateral faults of the Tien Shan and Pamir (in Russian), *in* Faults and horizontal movements of the earth's crust: Moscow, Nauka, Trudy Geological Institute, Akademi Nauk, USSR, 80, p. 152–172.

Butler, R.W.H.,1986, Thrust tectonics, deep structure and crustal subduction in the Alps and Himalayas: Journal of the The Geological Society of London, v. 143, p. 857–873.

Butler, R.W.H., Matthews, S. J., and Parish, M.,1986, The NW external Alpine thrust belt and its implications for the geometry of the western Alpine orogen, *in* Coward, M. P., and Ries, A. C., eds., Collision Tectonics: Geological Society of London Special Publication 19, p. 245–260.

Chang, Cheng-fa and Cheng Hsi-lan, 1973, Some tectonic features of the Mt. Jolmo Lungma area, southern Tibet: China, Scientia Sinica, v. 16, p. 257–265.

Chang Cheng-fa and Pan Yu-shen, 1981, A brief discussion on the tectonic evolution of the Qinghai-Xizang Plateau, *in* Geological and ecological studies of Qinhai-Xizang Plateau, Volume I: Geology, geological history and origin of Qinghai-Xizang Plateau: Beijing, Science Press, p. 1–18.

Chatelain, J.-L., Roecker, S. W., Hatzfeld, D., and Molnar, P., 1980, Microearthquake seismicity and fault plane solutions in the Hindu-Kush region and their tectonic implications: Journal of Geophysical Research, v. 85, p. 1365–1387.

Chen, W.-P., and Molnar, P., 1981, Constraints on the seismic wave velocity structure beneath the Tibetan Plateau and their tectonic implications: Journal of Geophysical Research, v. 86, p. 5937–5962.

—— , 1983, Focal depths of intracontinental and intraplate earthquakes and their implications for the thermal and mechanical properties of the lithosphere: Journal of Geophysical Research, v. 88, p. 4183–4214.

Chen, W.-P., Nabelek, J. L., Fitch, T. J., and Molnar, P., 1981, An intermediate depth earthquake beneath Tibet: Source characteristics of the event of September 14, 1976: Journal of Geophysical Research, v. 86, p. 2863–2876.

Chen Yan, Cogné, J.-P., and Courtillot, V., 1993, New Cretaceous paleomagnetic poles from the Tarim Basin, northwestern China: Earth and Planetary Science Letters, v. 114, p. 17–38.

Chopin, C., 1987, Very-high-pressure metamorphism in the western Alps: Implications for subduction of continental crust: Philosophical Transactions of the Royal Society of London, series A, v. 321, p. 183–195.

Coward, M. P., Butler, R.W.H., Asif, K. M., and Knipe, R. J., 1987, The tectonic history of Kohistan and its implications for Himalayan structure: Journal of the Geological Society of London, v. 144, p. 377–391.

Davidson, R. M., Kraidenkov, G. P., and Salibaev, G. Kh., 1982, Stratigraphy of Paleogene deposits of the Tadjik Depression and adjacent territories (in Russian): Dushanbe, Donish, 151 p.

Davis, G. A., and Burchfiel, B. C., 1973, Garlock fault: An intracontinental transform structure, southern California: Geological Society of America Bulletin, v. 84, p. 1407–1422.

DeMets, C., Gordon, R. G., Argus, D. F., and Stein, S., 1990, Current plate motions: Geophysical Journal International, v. 101, p. 425–478.

Desio, A., 1965, Sulla struttura tettonica dell'Asia centrale: Accademia Nazionale dei Lincei, Estratto dai rendiconti della classe de scienze fisiche, matemiche e naturali, Series 8, v. 38, p. 780–786.

Desio, A., ed., 1975, Geology of Central Badakhshan (north-east Afghanistan): Italian expeditions to the Karakorum (K2) and Hindu Kush, III: Geology-petrology, v. 3, Leiden, E. J. Brill, 628 p.

Dronov, V. I., ed., 1980, Geology and helpful fossils from Afghanistan, part 1 (in Russian): Moscow, Nedra, 536 p.

Dronov, V. I., 1986, Stratigraphy of the Jurassic and older volcanogenic sedimentary sequence in the Bashgumbez, Irikyak, Sedek, and Tashdzhilga river basins (Southeastern Pamir) (in Russian): Doklady Akademi Nauk Tadjik SSR, v. 29, p. 549–553.

—— , 1988, The volcanogenic type of Triassic deposits in Southeastern Pamir (in Russian): Doklady Akademi Nauk USSR, v. 303, p. 437–440.

Dzhalilov, M. R., 1971, Stratigraphy of the upper Cretaceous deposits of the Tadjik Depression (in Russian): Dushanbe, Donish, 210 p.

Dziewonski, A. M., Ekström, G., Woodhouse, J. H., and Zwart, G., 1989, Centroid-moment tensor solutions for April-June 1988: Physics of Earth and Planetary Interiors, v. 54, p. 199–209.

Ekström, G. A., 1987, A broad band method of earthquake analysis [Ph.D. thesis]: Cambridge, Harvard University, 216 p.

England, P. C., and Houseman, G. A., 1989, Extension during continental convergence, with application to the Tibetan Plateau: Journal of Geophysical Research, v. 94, p. 17561–17579.

Feng Xianyue, Yang Zhang, Luan Chaoqun, Yu Dexuan, Kou Dabing, and

Zhang Yong, 1986, The Wuqia earthquake of Xinjiang (in Chinese): Earthquake Research in China, v. 2, p. 56–60.

Fleitout, L., and Froidevaux, C., 1982, Tectonics and topography for a lithosphere containing density heterogeneieties: Tectonics, v. 1, p. 21–56.

Fredrich, J., McCaffrey, R., and Denham, D., 1988, Source parameters of seven large Australian earthquakes determined by body waveform inversion: Geophysical Journal, v. 95, p. 1–13.

Gaetani, M., and six others, 1990, The north Karakorum side of the Central Asia geopuzzle: Geological Society of America Bulletin, v. 102, p. 54–62.

Gansser, A., 1964, The geology of the Himalayas: New York, Wiley-Interscience, 289 p.

——, 1979, Reconnaissance visit to the ophiolites in Baluchistan and the Himalaya, *in* Farah, A., and DeJong, K. A., eds., Geodynamics of Pakistan: Quetta, Geological Survey of Pakistan, p. 193–213.

——, 1980, The significance of the Himalayan suture zone: Tectonophysics, v. 62, p. 37–52.

Geological map of Afghanistan, 1977, scale 1:500,000, ed. by V. I. Dronov.

Geological map of the Tadjik SSR and adjoining territories, 1984, scale 1:500,000, ed. by N. G. Vlasov and Yu. A. D'yakov.

Geological map of the Tibetan Plateau, 1980, scale 1:1,500,000, Beijing, China.

Geological map of the Xinjiang-Uygur Autonomous Region, China, 1985, scale 1:2,000,000, Beijing.

Girardeau, J., and seven others, 1984, Tectonic environment and geodynamic significance of the neo-Cimmerian Dongqiao ophiolite, Banggong-Nujiang suture zone: Nature, v. 307, p. 27–31.

Girardeau, J., Marcoux, J., and Montenat, C., 1989, The neo-Cimmerian ophiolite belt in Afghanistan and Tibet, *in* Sengör, A.M.C., ed., Tectonic evolution of the Tethyan region: Dordrecht, Kluwer, p. 477–504.

Grandjacquet, C., and Le Pichon, X., 1987, Lower crust subduction in the western Alps? [abs.]: Terra Cognita, v. 7, p. 152.

Gubin, I. Ye., 1960, Regularities of seismic phenomena in the Territory of Tadjikistan (in Russian): Moscow, Academi Nauk USSR, 464 p.

Guseva, T. G., 1986, Contemporary movements of the earth's crust in the transition zone from the Pamir to the Tien Shan (in Russian): Moscow, Institute of Physics of the Earth, Academi Nauk, 171 p.

Guseva, T. V., Lukk, A. A., Pevnev, A. K., Skovorodkin, Yu. P., and Shevchenko, V. I., 1983, Geodynamics of the Garm polygon region in Tadjikistan: Izvestiya Akademi Nauk USSR, Earth Physics, v. 19 (English translation), p. 506–518.

Gutenberg, B., and Richter, C. F., 1954, Seismicity of the earth and associated phenomena: New York, Hafner, 310 p.

Hamburger, M. W., Sarewitz, D. R., Pavlis, T. L., and Popandopulo, G. A., 1992, Structural and seismic evidence for intracontinental subduction in the Peter the First Range, Central Asia: Geological Society of America Bulletin, v. 104, p. 397–408.

Harris, N.B.W., Xu Ronghua, Lewis, C. L., and Jin Chengwei, 1988a, Plutonic rocks of the 1985 Tibet geotraverse, Lhasa to Golmud: Philosophical Transactions of the Royal Society of London, series A, v. 327, p. 145–168.

Harris, N.B.W., Xu Ronghua, Lewis, C. L., Hawkesorth, C. J., and Zhang Uyquan, 1988b, Isotope geochemistry of the 1985 Tibet geotraverse, Lhasa to Golmud: Philosophical Transactions of the Royal Society of London series A, v. 327, p. 263–285.

Hayden, H. H., 1915, Notes on the geology of Chitral, Gilgit, and the Pamirs: Records of the Geological Survey of India, v. 45, pt. 4, p. 1–30.

Helwig, J., 1976, Shortening of continental crust in orogenic belts and plate tectonics: Nature, v. 260, p. 768–770.

Hendrix, M. S., and six others, 1992, Sedimentary record and climatic implications of recurrent deformation in the Tien Shan: Evidence from Mesozoic strata of the north Tarim, south Junggar, and Turpan basins, northwest China: Geological Society of America Bulletin, v. 104, p. 53–79.

Holt, W. E., and Wallace, T. C., 1990, Crustal thickness and upper mantle velocities in the Tibetan Plateau region from the inversion of regional Pnl waveforms: Evidence for a thick upper mantle lid beneath southern Tibet: Journal of Geophysical Research, v. 95, p. 12499–12526.

Isacks, B. J., Oliver, J., and Sykes, L. R., 1968, Seismology and the new global tectonics: Journal of Geophysical Research, v. 73, p. 5855–5899.

Isacks, B. L., and Barazangi, M., 1977, Geometry of Benioff zones: Lateral segmentation and downwards bending of the subducted lithosphere, *in* Twalwani, M., and Pitman, W. C. III, eds., Island arcs, deep sea trenches, and back-arc basins: Maurice Ewing Series 1, Washington, D.C., American Geophysical Union, p. 99–114.

Jackson, J., Molnar, P., Patton, H., and Fitch, T. J., 1979, Seismotectonic aspects of the Markansu Valley, Tadjikstan, earthquake of August 11, 1974: Journal of Geophysical Research, v. 84, p. 6157–6167.

James, D. E., 1971, Andean crustal and upper mantle structure: Journal of Geophysical Research, v. 76, p. 3246–3271.

Johnson, G. D., Raynolds, R.G.H., and Burbank, D. W., 1986, Late Cenozoic tectonics and sedimentation in the north-west Himalayan foredeep: I. Thrust ramping and associated deformation in the Potwar region, *in* Allen, P. A., and Homewood, P., eds., Foreland Basins: International Association of Sedimentology Special Publication 8, Oxford, Blackwell, p. 273–291.

Karapetov, S. S., Sonin, I. I., and Khain, V. Ye, 1975, On several important peculiarities of the structure and development of the Afghan-Pamir segment of the Alpine fold belt of Eurasia (in Russian): Bulletin of Moscow University, Geology, no. 3, p. 38–46.

Kariev, A., 1977, Early Cretaceous sedimentary accumulations and the geologic development of the mountainous framework of the Tadjik Depression (in Russian): Izvestiya, Akademi Nauk, USSR, Geological Series, p. 115–125.

Kazmin, V. G., and Faradzhev, V. A.,1961, Tectonic development of the Yarkand sector of the Kunlun (in Russian): Soviet Geology, no. 8, p. 45–57.

Khalturin, V. I., Rautian, T. G., and Molnar, P., 1977, The spectral content of Pamir–Hindu Kush intermediate depth earthquakes: Evidence for a high-Q zone in the upper mantle: Journal of Geophysical Research, v. 82, p. 2931–2943.

Kolchanov, V. P., Kulyakov, V. V., Mikhailov, K. Ya., and Pashkov, B. R., 1971, New data on the stratigraphy of the Precambrian and Paleozoic formations on the northern front of the western Hindu Kush (in Russian): Soviet Geology, no. 3, p. 130–136.

Konopaltsev, I. M., 1971, Measurements of crustal movements in the Garm area, 1948–1970 (in Russian): Geotektonika, no. 5, p. 111–116.

Kosminskaya, I. P., Belyaevsky, N. A., and Volvovsky, I. S., 1969, Explosion seismology in the USSR, *in* Hart, P. J., ed., The earth's crust and upper mantle: Geophysical Monograph 13, Washington, D.C., American Geophysical Union, p. 195–208.

Kostrov, V. V.,1974, Seismic moment, and energy of earthquakes, and the seismic flow of rock (in Russian): Izvestiya, Akademi Nauk, USSR, Physics of the solid earth 1, p. 23–44.

Kuchai, V.K., and Trifonov, V. G., 1977, Young left-lateral strike slip along the zone of the Darvaz-Karakul' fault (in Russian): Geotektonika, no. 3, p. 91–105.

Kuchai, V. K., Pevnev, A. K., and Guseva, T. V., 1978, On the nature of contemporary tectonic movements in the Vakhsh overthrust zone (in Russian): Doklady Akademi Nauk USSR, v. 240, p. 673–678.

Kuchai, V. K., Guseva, T. V., and Ulashina, S. A., 1981, On the geodynamics of faults (in Russian): Izvestiya, Akademi Nauk, USSR. Geology Series, p. 45–56.

Kulagina, M. V., Lukk, A. A., and Kulagin, B. K., 1974, Block structure of the earth's crust of Tadjikstan (in Russian), *in* Searches for precursors of earthquakes in prediction polygons: Moscow, Nauka, p. 70–84.

Langston, C. A., and Dermengian, J. M., 1981, Comment *on* 'Seismotectonic aspects of the Markansu Valley, Tadjikstan, earthquake of August 11,

1974' by James Jackson, Peter Molnar, Howard Patton, and Thomas J. Fitch: Journal of Geophysical Research, v. 86, p. 1091–1093.

Langston, C. A., and Helmberger, D. V., 1975, A procedure for modeling shallow dislocation sources: Geophysical Journal of the Royal Astronomical Society, v. 42, p. 117–130.

Laubscher, H., 1988, Material balance in Alpine orogeny: Geological Society of America Bulletin, v. 100, p. 1313–1328.

Legler, B. A., and Przhiyalgovskaya, I. A., 1979, The interaction of the Indian and Asian lithospheric plates and tectonics of the Tadjik Depression (in Russian), *in* Structure of lithospheric plates: Institute of Oceanology, Akademi Nauk, USSR, Moscow, p. 125–188.

Leith, W., 1982, Rock assemblages in Central Asia and the evolution of the southern Asian margin: Tectonics, v. 1, p. 303–318.

—— , 1985, A mid-Mesozoic extension across Central Asia?: Nature, v. 313, p. 567–570.

Leith, W., and Alvarez, W., 1985, Structure of the Vakhsh fold-and-thrust belt, Tadjik SSR: Geologic mapping on a Landsat image base: Geological Society of America Bulletin, v. 96, p. 875–885.

Leith, W., and Simpson, D. W., 1986, Seismic domains within the Gissar-Kokshal seismic zone, Soviet Central Asia: Journal of Geophysical Research, v. 91, p. 689–699.

Leith, W., Simpson, D. W., and Alvarez, W., 1981, Structure and permeability: Geologic controls on induced seismicity at Nurek reservoir, Tadjikistan, USSR: Geology, v. 9, p. 440–444.

Leonov, N. N., 1961, Tectonics and seismicity of the Pamir-Alai zone (in Russian): Moscow, Akademi Nauk, USSR, 164 p.

Le Pichon, X., Bergerat, F., and Roulet, M.-J., 1988, Plate kinematics and tectonics leading to the Alpine belt formation: A new analysis, *in* Clark, S. P., Burchfiel, B. C., and Suppe, J., eds., Processes in continental lithospheric deformation: Geological Society of America Special Paper 218, p. 111–131.

Leven, E. Ya., 1981, The age of Paleozoic volcanogenic formations of the Northern Pamir (in Russian): Izvestiya, Akademi Nauk, USSR, Geology Series, 9, p. 137–140.

Leven, E. Ya., and Shcherbovich, S. F., 1978, Fusilinids and stratigraphy of the Assel Stage of Darvaz (in Russian): Moscow, Nauka, 163 p.

Lillie, R. J., Johnson, G. D., Yousuf, M., Zamin, A.S.H., and Yeats, R. S., 1987, Structural development within the Himalayan foreland fold-and-thrust belt of Pakistan, *in* Beaumont, C., and Tankard, A. J., eds., Sedimentary basins and basin-forming mechanisms: Canadian Society of Petroleum Geologists Memoir 12, P. 379–392.

Liu, Q., Tapponnier, P., Boujot, L., and Zhang, Q., 1992, Recent movement along the northern segment of the Karkorum fault system, in the Multi-Tashgorgan graben, in International Symposium on the Karakorum and Kunlun Mountains, June 5–9, 1992, Kashi [sic], China, Abstracts, p. 92.

Lukk, A. A., and Nersesov, I. L., 1970, The deep Pamir–Hindu Kush earthquakes (in Russian), *in* Earthquakes in the USSR in 1966: Moscow, Nauka, p. 118–136.

Lukk, A. A., and Shevchenko, V. I., 1986, Character of the deformation of the earth's crust in the Garm region (Tadjikistan) from geological and seismological data: Izvestiya, Earth Physics (English translation), v. 22, p. 527–539.

Lukk, A. A., and Yunga, S. L., 1988, Geodynamics and the compressive deformation state of the lithosphere of Middle Asia (in Russian): Dushanbe, Donish, 280 p.

Malamud, A. C., 1973, Several regularities of the spatial distribution of Pamir–Hindu Kush deep focus earthquakes (in Russian): Izvestiya, Akademi Nauk, Tadjik SSR, Otdel Fiziki Matematicheskikh Geolicheskii i Khimii, v. 4, p. 70–73.

Matte, P., Tapponnier, P., Bourjot, L., Pan, Y., and Wang, Y., 1992, Tectonics in western Tibet, from the Kunlun to the Karakorum, *in* International Symposium on the Karakorum and Kunlun Mountains, June 5–9, 1992, Kashi [sic], China, Abstracts, p. 36.

McCaffrey, R., and Abers, G., 1988, SYN3: A program for inversion of teleseismic body wave forms on microcomputers: Air Force Geophysics Laboratory Technical Report AFGL-TR-88-0099, Hanscomb Air Force Base, Massachusetts, 50 p.

McKenzie, D. P., 1969, Speculations on the consequences and causes of plate motions: Geophysical Journal of the Royal Astronomical Society, v. 18, p. 1–32.

McKenzie, D., and O'Nions, R. K., 1983, Mantle reservoirs and ocean island basalts: Nature, v. 301, p. 229–231.

Mellors, R. J., Al-Shuki, H. J., Hamburger, M. W., Lukk, A. A., and Popandopulo, G. A., 1991, Velocity structure of the crust and upper mantle in the Pamir–Tien Shan region, Soviet Central Asia [abs.]: EOS Transactions of the American Geophysical Union, v. 72, p. 203.

Ménard, G., and Molnar, P., 1988, Collapse of a Hercynian Tibetan Plateau into a late Palaeozoic Basin and Range Province in southern Europe: Nature, v. 334, p. 235–237.

Ménard, G., Molnar, P., and Platt, J. P., 1991, Budget of crustal shortening and subduction of continental crust in the Alps: Tectonics, v. 10, p. 231–244.

Minakova, N. Ye., Sotiriadi, K. A., Egamberdyev, M., Nasrettdinov, Z. Z., Khikmatyllaev, B. C., and Talipov, C., 1975, Lithology, stratigraphy, geochemistry, and the presence of oil and gas in Paleogene deposits of the Syrkhandaryn Basin (in Russian): Tashkent, Fan, 153 p.

Mitronovas, W., and Isacks, B. L., 1971, Seismic velocity anomalies in the upper mantle beneath the Tonga-Kermadec island arc: Journal of Geophysical Research, v. 76, p. 7154–7180.

Molnar, P., and Chen, W. -P., 1983, Focal depths and fault plane solutions of earthquakes under the Tibetan plateau: Journal of Geophysical Research, v. 88, p. 1180–1196.

Molnar, P., and England, P., 1990, Late Cenozoic uplift of mountain ranges and global climate change: Chicken or egg?: Nature, v. 346, p. 29–34.

Molnar, P., and Gray, D., 1979, Subduction of continental lithosphere: Some constraints and uncertainties: Geology, v. 7, p. 58–62.

Molnar, P., and Jackson, J., 1981, Reply: Journal of Geophysical Research, v. 86, p. 1095–1096.

Molnar, P., and Lyon-Caen, H., 1989, Fault plane solutions of earthquakes and active tectonics of the northern and eastern parts of the Tibetan Plateau: Geophysical Journal International, v. 99, p. 123–153.

Molnar, P., and Tapponnier, P., 1975, Cenozoic tectonics of Asia: Effects of a continental collision: Science, v. 189, p. 419–426.

—— , 1978, Active tectonics of Tibet: Journal of Geophysical Research, v. 83, p. 5361–5375.

Molnar, P., Freedman, D., and Shih, J.S.F., 1979, Lengths of intermediate and deep seismic zones and temperatures in downgoing slabs of lithosphere: Geophysical Journal of the Royal Astronomical Society, v. 56, p. 41–54.

Molnar, P., Burchfiel, B. C., Liang K'uangyi, and Zhao Ziyun, 1987, Geomorphic evidence for active faulting in the Altyn Tagh and northern Tibet and qualitative estimates of its contribution to the convergence of India and Eurasia: Geology, v. 15, p. 249–253.

Montenat, Ch., Girardeau, J., and Marcoux, J., 1986a, La ceinture ophiolitique Néo-Cimmérienne au Tibet, dans les Pamirs et en Afghanistan: Evolution géodynamique comparative, *in* LeFort, P., Colchen, M., and Montenat, Ch., eds., Mémoires Science de la Terre No. 47, Evolution des domaines orogéniques d'Asie méridionale (de la Turquie à l'Indonésie): Nancy, France, Fondation Scientifique de la Géologie et de ses Applications, p. 229–252.

Montenat, Ch., Vachard, D., Ouedraogo, A., Bébien, J., and Desmet, A., 1986b, Le bassin volcano-sédimentaire de Kandahar (Afghanistan): Un jalon de la marge active nord-téthysienne ceinture au Malm et au Crétacé inférieure, *in* LeFort, P., Colchen, M., and Montenat, Ch., eds., Mémoires Science de la Terre No. 47, Evolution des domaines orogéniques d'Asie méridionale (de la Turquie à l'Indonésie): Nancy, France, Fondation Scientifique de la Géologie et de ses Applications, p. 253–283.

Nábělek, J., 1984, Determination of earthquake source parameters from inversion of body waves [Ph.D. thesis]: Cambridge, Massachusetts Institute of Technology, 361 p.

Nalivkin, D. V., 1915, Preliminary report on the expedition in the summer of 1915 to mountainous Bukhara and the western Pamir (in Russian): Izvestiya, Russian Geographical Society, v. 52, p. 12–18.

—— , 1926, Outline of the geology of Turkestan (in Russian): Moscow-Tashkent, Turkpechat', 184 p.

Nelson, M. R., McCaffrey, R., and Molnar, P., 1987, Source parameters for 11 earthquakes in the Tien Shan, Central Asia, determined by P and SH waveform inversion: Journal of Geophysical Research, v. 92, p. 12629–12648.

Ni, J., 1978, Contemporary tectonics in the Tien Shan region: Earth and Planetary Science Letters, v. 41, p. 347–355.

Nikonov, A. A., 1970, Differential analysis of the Quaternary tectonics of the Afghan-Tadjik Depression (in Russian): Geotektonika, no. 1, p. 101–107.

—— , 1974, Contemporary and Holocene seismotectonic dislocations in the South Tien Shan seismically active zone (in Russian): Izvestiya, Akademi Nauk USSR, Earth Physics, no. 12, p. 71–76.

—— , 1975, An analysis of tectonic movements along the Hindu Kush–Darvaz–Karakul fault zone in late Pliocene and Quaternary time (in Russian): Bulletin MOIP (of the Moscow Society for the investigation of Nature), Geology Section, v. 50, p. 5–23.

—— , 1977, Holocene and contemporary movements of the earth's crust (in Russian): Moscow, Nauka, 240 p.

Nikonov, A. A., Vakov, A. V., and Veselov, I. A., 1983, Seismotectonics and earthquakes in the converegent zone between the Pamir and Tien Shan (in Russian): Moscow, Nauka, 240 p.

Norin, E., 1946, Geological explorations in Western Tibet: Reports from the scientific expedition to the northwestern provinces of China under the leadership of Dr. Sven Hedin, Publication 29, (III), Geology 7: Thule, Stockholm, Tryckeri Aktiebolaget, 214 p.

Nowroozi, A. A., 1971, Seismo-tectonics of the Persian plateau, eastern Turkey, Caucasus, and Hindu Kush regions: Bulletin of the Seismological society of America, v. 61, p. 317–341.

Oliver, J., and Isacks, B., 1967, Deep earthquake zones, anomalous structures in the upper mantle, and the lithosphere: Journal of Geophysical Research, v. 72, p. 4259–4275.

Pan Yusheng, ed., 1992, Introduction to the integrated scientific investigation on Karakorum and Kunlun Mountains: Beijing, China Meteorological Press, 92 p.

Pashkov, B. R., and Budanov, V. I., 1990, The tectonics of the zone of intersection between the Southeastern and southwestern Pamir (in Russian): Geotektonika, no. 3, p. 70–79.

Pashkov, B. R., and Shvol'man, V. A., 1979, Riftogenic margins of Tethys in the Pamir (in Russian): Geotektonika, no. 6, p. 42–57.

Pashkov, B. R., Pyzh'yanov, I. V., and Belov, A. A., 1981, Paleozoic deposits of the autochthon and covering mass of the Central Pamir (in Russian): Izvestiya, Akademi Nauk, USSR, Geology Series 11, p. 93–105.

Pearce, J. A., and Deng Wanming, 1988, The ophiolites of the Tibetan geotraverses, Lhasa to Golmud (1985) and Lhasa to Kathmandu (1986): Philosophical Transactions of the Royal Society of London, series A, v. 327, p. 215–238.

Pearson, G. W., Pilcher, J. R., Baillie, M.G.L., Corbett, D. M., and Qua, F., 1986, High-precision ^{14}C measurement of Irish oaks to show the natural ^{14}C variations from AD 1840–5210 BC: Radiocarbon, v. 28, p. 911–934.

Peive, A. V., Burtman, V. S., Ruzhentsev, S. V., and Suvorov, A. I., 1964, Tectonics of the Pamir-Himalayan sector of Asia, *in* Report of the Twenty-Second Session, India, International Geological Congress: Part XI, Himalayan and Alpine Orogeny: New Delhi, International Geological Congress, p. 441–464

Platt, J. P., 1986, Dynamics of orogenic wedges and the uplift of high-pressure metamorphic rocks: Geological Society of America Bulletin, v. 98, p. 1037–1053.

Pospelov, I. I., 1987, Formations and tectonic development of the late Variscides of the South Tien Shan and Northern Pamir (in Russian), *in* Pushcharov, Yu. M., and Khvorova, I. V., eds., Early geosynclinal formations and structures: Moscow, Nauka, p. 149–178.

Pospelov, I. I., and Sigachev, S. P., 1987, On retrocharriage in the Pamir (in Russian): Doklady Akadmi Nauk USSR, no. 3, p. 678–683.

Pozzi, J.-P., and Feinberg, H., 1991, Paleomagnetism in the Tadjikistan: Continental shortening of the European margin in the Pamirs during Indian Eurasian collision: Earth and Planetary Science Letters, v. 103, p. 365–378.

Pyzh'yanov, I. V., and Sonin, I. I., 1977, Basic features of the stratigraphy of the Upper Paleozoic and Lower Mesozoic of Afghanistan (in Russian): Izvestiya of Universities (Vuz), Geology and Prospecting 12, p. 30–39.

Pyzh'yanov, I. V., Dronov, V. I., Karapetov, S. S., and Sonin, I. I., 1978, The stratigraphy of Carboniferous deposits of Afghanistan (in Russian): Bulletin MOIP (of the Moscow Society for the Investigation of Nature), Geology Section, v. 53, p. 78–91.

Richardson, S. W., and England, P. C., 1979, Metamorphic consequences of crustal eclogite production in orogenic overthrust zones: Earth and Planetary Science Letters, v. 42, p. 183–190.

Roecker, S. W., 1982, Velocity structure of the Pamir–Hindu Rush region: Possible evidence for subducted crust: Journal of Geophysical Research, v. 87, p. 945–959.

Roecker, S. W., and six others, 1980, Seismicity and fault plane solutions of intermediate depth earthquakes in the Pamir–Hindu Kush region: Journal of Geophysical Research, v. 85, p. 1358–1364.

Roecker, S. W., Tucker, B., King, J., and Hatzfeld, D., 1982, Estimates of Q in Central Asia as a function of frequency and depth using the coda of locally recorded earthquakes: Bulletin of the Seismological Society of America, v. 72, p. 129–149.

Roecker, S. W., Yeh, Y. H., and Tsai, Y. B., 1987, Three dimensional P and S wave structures beneath Taiwan: Deep structure beneath an arc-continent collision: Journal of Geophysical Research, v. 92, p. 10547–10570.

Romanowicz, B, 1982, Constraints on the structure of the Tibet Plateau from pure path phase velocities of Love and Rayleigh waves: Journal of Geophysical Research, v. 87, p. 6865–6883.

Ruzhentsev, S. V.,1963, Strike-slip faults of the Southeastern Pamir (in Russian), *in* Faults and horizontal movements of the earth's crust: Moscow, Nauka, Trudy Geological Institute, Akademi Nauk, USSR, 80, p. 113–127.

—— , 1968, Tectonic development of the southeastern Pamir and the role of Horizontal movements in the formation of its alpine structure (in Russian): Moscow, Nauka, 202 p.

—— , 1971, Peculiarities of the structure and the mechanism of formation detached covers (in Russian): Moscow, Nauka, 136 p.

—— , 1990a, The study and mapping of cover [thin-skinned] folds (in Russian), *in* Pushcharovskii, Yu. M., and Trifonov, V. G., eds., Tectonic stratification of the lithosphere and regional geologic investigations: Moscow, Nauka, p. 72–82.

—— , 1990b, The Pamir (in Russian), *in* Pushcharovskii, Yu. M., and Trifonov, V. G., eds., Tectonic stratification of the lithosphere and regional geologic investigations: Moscow, Nauka, p. 214–225.

Ruzhentsev, S. V., and Shvol'man, V. A., 1981a, Tectonic zoning of the Pamirs and Afghanistan, *in* Sinha, A. K., ed., Contemporary geoscientific researches in Himalaya: Dehra Dun, India, Bishen Singh Mahendra Pal Singh, p. 53–59.

—— , 1981b, Tectonics and structure of the Pamir metamorphics, *in* Saklani, P. S., ed., Metamorphic tectonites of the Himalaya: New Delhi, Today and Tomorrow's Printers and Publishers, p. 27–41.

—— , 1982, The Pamirs, *in* Mahel, M., ed., Alpine structural elements: Carpathian-Balkan-Caucasus-Pamir orogenic zone: Bratislava, VEDA, p. 115–130.

Ruzhentsev, S. V., Pospelov, I. I., and Sukhov, A. N.,1977, The tectonics of the Kalaikhumb-Sauksai zone of the Northern Pamir (in Russian): Geotektonika, no. 4, p. 68–80.

Ruzhentsev, S. V., Shvol'man, V. A., and Pashkov, B. R., 1983, The tectonic development of the Pamir-Himalayan sector of the Alpine fold belt (in Russian), in Gubin, I. Ye., and Zakharov, S. A., eds., The tectonics of the Tien Shan and Pamir: Moscow, Nauka, p. 167–175.

Searle, M. P., 1991, Geology and tectonics of the Karakoram Mountains: Chichester, J. Wiley & Sons, 358 p.

Sengör, A.M.C., 1979, Mid-Mesozoic closure of Permo-Triassic Tethys and its implications: Nature, v. 279, p. 590–593.

——, 1984, The Cimmeride orogenic system and the tectonics of Eurasia: Geological Society of America Special Paper 195, 82 p.

Sengör, A.M.C., and Hsü, K. J., 1984, The Cimmerides of eastern Asia: History of the eastern end of palaeo-Tethys: Societé Géologique de la France Mémoir 147, p. 139–167.

Sengör, A.M.C., Altiner, D., Cin, A., Ustaömer, T., and Hsü, K. J., 1988, Origin and assembly of the Tethyside orogenic collage at the expense of Gondwana Land, *in* Audley-Charles, M. G., and Hallam, A., Gondwana and Tethys: Geological Society Special Publication 37, p. 119–181.

Shvol'man, V. A., 1977, Tectonic development of the Pamir in the Cretaceous and Paleogene periods (in Russian): Transactions of the Geological Institute, Akademi Nauk USSR, v. 302, Moscow, Nauka, 160 p.

——, 1978, Relicts of the Mesotethys in the Pamirs: Himalayan Geology, v. 8, Part 1, p. 369–378.

——, 1980, A Mesozoic ophiolitic complex in the Pamir (in Russian): Geotektonika, no. 6, p. 72–82.

Shvol'man, V. A., and Pashkov, B. R., 1986, Early Cretaceous tectonic zonations of Central Asia (in Russian): Doklady Akademi Nauk USSR, v. 286, p. 951–954.

Sinitsyn, V. M., 1957, The northwest part of the Tarim Basin (in Russian): Moscow, Akademi Nauk USSR, 250 p.

——, 1959, Central Asia (in Russian): Moscow, Gozgeografgiz, 456 p.

Skobelev, S. F., 1977, Horizontal compression and development of folds in the Peter the First Range (in Russian): Geotektonika, no. 2, p. 105–119.

Skobelev, S. F., Trifonov, V. G., and Vostrikov, G. A., 1988, The Pamir-Himalayan region of disharmonic piling up of continental lithosphere (in Russian), *in* Kropotkin, P. N., ed., Neotectonics and contemporary geodynamics of mobile belts: Moscow, Nauka, p. 188–233.

Soboleva, O. V., Kinyapina, T. A., and Shklyar, G. P., 1975, Focal mechanisms of strong earthquakes in the central part of the Tadjik Depression (in Russian), *in* Strong earthquakes in Middle Asia and Kazakhstan: Dushanbe, Donish, p. 88–105.

Stöcklin, J., 1977, Structural correlation of the Alpine ranges between Iran and Central Asia, *in* Livre à la mémoire de Albert F. de Lapparent, consacré aux recherches géologiques dans les Chaines Alpines de l'Asie du Sud-ouest: Societé Géologique de la France, Mémoire hors-série 8, p. 333–353.

Suess, E., 1904, The face of the earth, v. I (English translation): Oxford, Oxford University Press, 604 p.

Suvorov, A. I., 1968, Regularities of the structure and formation of deep faults (in Russian): Moscow, Nauka, 316 p.

Tahirkheli, R.A.K., Mattauer, M., Proust, F., and Tapponnier, P., 1979, The India-Eurasia suture zone in northern Pakistan: Synthesis and interpretation of recent data at a plate scale, in Farah, A., and DeJong, K. A., eds., Geodynamics of Pakistan: Geological Survey of Pakistan, p. 125–230.

Tapponnier, P., and Molnar, P., 1979, Active faulting and Cenozoic tectonics of the Tien Shan, Mongolia and Baykal Regions: Journal of Geophysical Research, v. 84, p. 3425–3459.

Tapponnier, P., Mattauer, M., Proust, F., and Cassaigneau, C., 1981, Mesozoic ophiolites, sutures, and large-scale tectonic movements in Afghanistan: Earth and Planetary Science Letters, v. 52, p. 355–371.

Tapponnier, P., Peltzer, G., and Armijo, R., 1986, On the mechanics of the collision between India and Asia, *in* Coward, M. P., and Ries, A. C., eds., Collision tectonics: Geological Society of London, p. 115–157.

Thakur, V. C., and Misra, D. K., 1984, Tectonic framework of the Indus and Shyok suture zones in eastern Ladakh, northwest Himalaya: Tectonophysics, v. 101, p. 207–220.

Trifonov, V. G., 1978, Late Quaternary tectonic movements of western and central Asia: Geological Society of America Bulletin, v. 89, p. 1059–1072.

——, 1983, Late Quaternary tectogenesis (in Russian): Moscow, Nauka, 224 p.

Varentsov, M. I., Aleshina, Z. I., and Kornienko, G. Ye., 1977, Tectonics and oil and gas of the South Tadjik basin (in Russian): Moscow, Nauka, 108 p.

Vinnik, L. P., and Lukk, A. A., 1973, Velocity inhomogeneities in the upper mantle of Central Asia (in russian): Doklady Akademi Nauk USSR, v. 213, p. 580–583.

——, 1974, Lateral inhomogeneities in the upper mantle under the Pamir–Hindu Kush (in Russian): Izvestiya, Akademi Nauk, USSR Physics of the solid earth, p. 9–22.

Vinnik, L. P., Lukk, A. A., and Nersesov, I. L., 1977, Nature of the intermediate seismic zone in the mantle of the Pamir–Hindu Kush: Tectonophysics, v. 38, p. T9–T14.

Vinnik, L. P., Lukk, A. A., and Mirzokurbanov, M., 1978, Quantitative analysis of inhomogeneities in the upper mantle of the Pamir–Hindu Kush (in Russian): Izvestiya, Akademi Nauk, USSR, Physics of the solid earth 5, p. 3–16.

Volvov'skii, I. S., and Volvov'skii, B. S., 1975, Cross sections of the earth's crust of the territory of the USSR from data of deep seismic sounding: Moscow, Soviet Radio, 268 p. (English translation: Addis Translations International, Portola Valley, California, 1978).

Wadia, D. N., 1975, Geology of India: New Delhi, Tata–McGraw-Hill, 508 p.

Wellman, H. W., 1966, Active wrench faults in Iran, Afghanistan and Pakistan: Geologische Rundschau, v. 53, p. 716–735.

Wernicke, B. P., Spencer, J. E., Burchfiel, B. C., and Guth, P. L., 1982, Magnitude of extension in the southern Great Basin: Geology, v. 10, p. 499–502.

Xu Ronghua, Schärer, U., and Allègre, C. J., 1985, Magmatism and metamorphism in the Lhasa block (Tibet): A geochemical study: Journal of Geology, v. 93, p. 41–57.

Xu Ronghua, Zhang Yuquan, Vidal, P., and Arnaud, N., 1992, Two plutonic belts in western Kunlun, *in* International Symposium on the Karakorum and Kunlun Mountains, June 5–9, 1992, Kashi [sic] China, Abstracts, p. 62.

Yermilin, B. I., and Chigarev, N. V., 1981, Mountain formation and seismicity of the Pamir-Alai (in Russian): Moscow, Nauka, 128 p.

Zakharov, S. A., 1948, On the east-west strike-slip faults on the south front of the Gissar Range (in Russian): Reports of the Tadjik Filial Akademi Nauk of the USSR, v. 5, Stalinabad, p. 3–5.

——, 1955, On the relationship of the Tadjik Depression and the Gissar Range (in Russian): Izvestiya, Branch of Natural Sciences of the Tadjik SSR, v. 9, Stalinabad, p. 15–20.

——, 1958, Sediment structure of the Mesozoic and Cenozoic of the Tadjik Depression (in Russian): Stalinabad, Donish, 230 p.

——, 1964, Cardinal questions of tectogenesis in connection with the direction of searches for oil and gas in the Tadjik Depression and the bases of seismotectonic regionalization of southern Tadjikistan (in Russian): *in* Problems of the geology of Tadjikistan, Dushanbe, Donish, p. 3–78.

——, 1967, On the nature of the Darvaz fault (in Russian): Doklady, Akademi Nauk USSR, v. 175, p. 893–896.

MANUSCRIPT ACCEPTED BY THE SOCIETY October 16, 1992

Printed in U.S.A.

Typeset by WESType Publishing Services, Inc., Boulder, Colorado
Printed in U.S.A. by Malloy Lithographing, Inc., Ann Arbor, Michigan